Anna Weigand

Relativistic Energy-consistent Pseudopotentials for f-Elements

Anna Weigand

Relativistic Energy-consistent Pseudopotentials for f-Elements

5f-in-core Pseudopotentials for Actinides and 4f-in-core Pseudopotentials for Tetravalent Lanthanides

Südwestdeutscher Verlag für Hochschulschriften

Impressum/Imprint (nur für Deutschland/ only for Germany)
Bibliografische Information der Deutschen Nationalbibliothek: Die Deutsche Nationalbibliothek verzeichnet diese Publikation in der Deutschen Nationalbibliografie; detaillierte bibliografische Daten sind im Internet über http://dnb.d-nb.de abrufbar.

Alle in diesem Buch genannten Marken und Produktnamen unterliegen warenzeichen-, marken- oder patentrechtlichem Schutz bzw. sind Warenzeichen oder eingetragene Warenzeichen der jeweiligen Inhaber. Die Wiedergabe von Marken, Produktnamen, Gebrauchsnamen, Handelsnamen, Warenbezeichnungen u.s.w. in diesem Werk berechtigt auch ohne besondere Kennzeichnung nicht zu der Annahme, dass solche Namen im Sinne der Warenzeichen- und Markenschutzgesetzgebung als frei zu betrachten wären und daher von jedermann benutzt werden dürften.

Verlag: Südwestdeutscher Verlag für Hochschulschriften Aktiengesellschaft & Co. KG
Dudweiler Landstr. 99, 66123 Saarbrücken, Deutschland
Telefon +49 681 37 20 271-1, Telefax +49 681 37 20 271-0
Email: info@svh-verlag.de
Zugl.: Köln, Universität, Diss., 2009

Herstellung in Deutschland:
Schaltungsdienst Lange o.H.G., Berlin
Books on Demand GmbH, Norderstedt
Reha GmbH, Saarbrücken
Amazon Distribution GmbH, Leipzig
ISBN: 978-3-8381-1710-2

Imprint (only for USA, GB)
Bibliographic information published by the Deutsche Nationalbibliothek: The Deutsche Nationalbibliothek lists this publication in the Deutsche Nationalbibliografie; detailed bibliographic data are available in the Internet at http://dnb.d-nb.de.

Any brand names and product names mentioned in this book are subject to trademark, brand or patent protection and are trademarks or registered trademarks of their respective holders. The use of brand names, product names, common names, trade names, product descriptions etc. even without a particular marking in this works is in no way to be construed to mean that such names may be regarded as unrestricted in respect of trademark and brand protection legislation and could thus be used by anyone.

Publisher: Südwestdeutscher Verlag für Hochschulschriften Aktiengesellschaft & Co. KG
Dudweiler Landstr. 99, 66123 Saarbrücken, Germany
Phone +49 681 37 20 271-1, Fax +49 681 37 20 271-0
Email: info@svh-verlag.de

Printed in the U.S.A.
Printed in the U.K. by (see last page)
ISBN: 978-3-8381-1710-2

Copyright © 2010 by the author and Südwestdeutscher Verlag für Hochschulschriften Aktiengesellschaft & Co. KG and licensors
All rights reserved. Saarbrücken 2010

Formeln sind so etwas wie immaterielle Raumsonden, mit denen wir ein Stückchen über die Grenze unseres Vorstellungsvermögens hinaus in Bereiche der Wirklichkeit vorstoßen können, die uns sonst verschlossen bleiben.

Hoimar von Ditfurth

Kurzzusammenfassung

Im Rahmen dieser Doktorarbeit wurden relativistische energie-konsistente Pseudopotentiale (PPs) für f-Elemente justiert. Bei der Pseudopotentialmethode werden nur die chemisch relevanten Valenzelektronen explizit in den Rechnungen behandelt und die wichtigsten relativistischen Effekte durch eine geeignete Parametrisierung implizit berücksichtigt. Deshalb werden PPs häufig zur theoretischen Untersuchung von f-Elementverbindungen verwendet, für deren Berechnung die hohe Anzahl an Elektronen und die großen relativistischen Effekte die Hauptherausforderungen darstellen. Außerdem können Schwierigkeiten aufgrund von offenen Schalen durch PPs vermieden werden, indem sie die offenen Schalen in den Rumpfbereich einbeziehen, wie es z.B. für die *f-in-core* PPs der Fall ist. Wenn allerdings die offene f-Schale nicht explizit behandelt wird, muss für jede Oxidationsstufe ein eigenes PP angepasst werden. Diese Doktorarbeit vervollständigt die bereits vorhandenen quasirelativistischen energie-konsistenten *f-in-core* PPs, d.h. für die Actinoide wurden zwei- (Pu–No), vier- (Th–Cf), fünf- (Pa–Am) und sechswertige (U–Am) *5f-in-core* PPs und für die Lanthanoide vierwertige (Ce–Nd, Tb, Dy) *4f-in-core* PPs justiert. Zu diesen PPs wurden polarisierte Valenz-Double-, Valenz-Triple- und Valenz-Quadruple-Zeta-Basissätze zur Anwendung in Molekülrechnungen optimiert, welche kleinere Basissätze umfassen, die zur Berechnung von Kristallen eingesetzt werden können. Zusätzlich wurden im Falle der zwei-, drei- und vierwertigen Actinoide Polarisationspotentiale zur Berücksichtigung der vernachlässigten statischen und dynamischen Rumpfpolarisation angepasst. Die atomaren Testrechnungen an den Ionisierungspotentialen der Actinoide und die molekularen Testrechnungen an den Actinoid- und Lanthanoidfluoriden auf Hartree-Fock- bzw. Coupled-Cluster-Niveau zeigen außer für PuF_2 und NpF_6–AmF_6 eine gute Übereinstimmung mit entsprechenden *f-in-valence* Pseudopotentialrechnungen bzw. experimentellen Werten. Für PuF_2 ist der Grund für die große Abweichung, dass die zweiwertige Oxidationsstufe für Plutonium nicht stabil ist. Für die sechswertigen PPs zeigt sich hingegen, dass die *5f-in-core* Näherung außer für Uran ($5f^0$) versagt. Da die *5f-in-core* PPs auch für Actinocene, Actinyl-Ionen und Uranyl(VI)-Komplexe erfolgreich angewendet wurden, sollte die *f-in-core* Näherung für Verbindungen, in denen die f-Orbitale nicht signifikant an der chemischen Bindung beteiligt sind, eine effiziente Rechenmethode darstellen.

Zusätzlich zu den quasirelativistischen *f-in-core* PPs wurde ein kürzlich justiertes *5f-in-valence* PP für Uran, welches sowohl skalar-relativistische Effekte als auch die Spin-Bahn-Kopplung berücksichtigt, zur Berechnung der U^{5+}- und U^{4+}-Feinstrukturaufspaltung verwendet. Diese Rechnungen ergaben zuverlässige Ergebnisse und bestätigen somit frühere Testrechnungen an Uranmonohydrid.

Abstract

In this thesis relativistic energy-consistent pseudopotentials (PPs) for f-elements have been adjusted. The PP approach restricts the explicit calculations to the chemically relevant valence electron system and implicitly includes relativistic effects by means of a simple parameterization. Thus, it is a commonly used approximation to study molecules containing f-elements, where the large number of electrons and the significant relativistic effects are the main obstacles. Even difficulties due to open shells can be avoided, if these are included in the core, as it is the case for f-in-core PPs. However, if the f shell is not treated explicitly, one PP for each oxidation state has to be adjusted. This thesis completes the already existing quasirelativistic f-in-core PPs, i.e. 5f-in-core PPs for di- (Pu–No), tetra- (Th–Cf), penta- (Pa–Am), and hexavalent (U–Am) actinides and 4f-in-core PPs for tetravalent (Ce–Nd, Tb, Dy) lanthanides are presented. Corresponding molecular basis sets of polarized valence double- to quadruple-zeta quality have been derived. Smaller basis sets suitable for crystal calculations form subsets of these basis sets. Furthermore, core-polarization potentials for di-, tri-, and tetravalent actinides have been adjusted to account for the neglect of static and dynamic core-polarization. Atomic test calculations on actinide ionization potentials as well as molecular test calculations on actinide and lanthanide fluorides using the Hartree–Fock and coupled cluster method show satisfactory agreement with calculations using f-in-valence PPs and experimental data, respectively, except for PuF_2 and NpF_6–AmF_6. While for PuF_2 the large deviations are due to the fact that for plutonium the divalent oxidation state is not stable, in the hexavalent case the 5f-in-core approximation seems to reach its limitations except for uranium ($5f^0$). Moreover, the 5f-in-core PPs are successfully applied to actinocenes, actinyl ions, and uranyl(VI) complexes. Thus, the f-in-core PPs should be an efficient computational tool for those compounds, where the f orbitals do not participate significantly in chemical bonding. In addition to the quasirelativistic f-in-core PPs, the recently adjusted 5f-in-valence uranium PP including scalar-relativistic effects as well as spin–orbit coupling have been tested by calculating the fine-structure splittings of U^{5+} and U^{4+}. These test calculations gave reliable results and thus confirm earlier benchmark calculations on uranium monohydride.

Contents

Kurzzusammenfassung	iii
Abstract	v
List of Abbreviations	ix
1 Introduction	**1**
2 Theory	**7**
2.1 Effective Core Potentials	7
2.1.1 Motivation	7
2.1.2 Approximations	8
2.1.3 Model Potential vs. Pseudopotential	11
2.1.4 Relativistic Treatment	12
2.1.5 Valence-only Model Hamiltonian	16
2.1.6 Energy Adjustment	18
2.2 Quantum Mechanical Methods	21
2.2.1 Hartree–Fock Method	22
2.2.2 Configuration Interaction	25
2.2.3 Multi-configuration Self-consistent Field Method	26
2.2.4 Coupled Cluster Theory	27
2.2.5 Second-order Møller–Plesset Perturbation Theory	28
2.2.6 Density Functional Theory	31
3 Results and Discussion	**35**
3.1 5f-in-core Pseudopotentials for Actinides	35
3.1.1 Adjustment of the Pseudopotentials	35

		3.1.2	Adjustment of the Core-polarization Potentials	37
		3.1.3	Optimization of the Valence Basis Sets	39
		3.1.4	Atomic Test Calculations	41
		3.1.5	Molecular Test Calculations	46
		3.1.6	Range of Applications	66
		3.1.7	Application to Actinocenes	67
		3.1.8	Application to Actinyl Ions	74
		3.1.9	Application to Uranyl(VI) Complexes	79
	3.2	4f-in-core Pseudopotentials for Lanthanides		99
		3.2.1	Adjustment of the Pseudopotentials	99
		3.2.2	Optimization of the Valence Basis Sets	100
		3.2.3	Molecular Test Calculations	101
	3.3	5f-in-valence Pseudopotential for Uranium		123
		3.3.1	Adjustment of the Pseudopotential	123
		3.3.2	Optimization of the Valence Basis set	124
		3.3.3	Atomic Test Calculations on U^{5+} and U^{4+}	125

4 Conclusion and Outlook   **143**

A Pseudopotentials   **149**

B Basis Sets   **161**

C Test Calculations and Applications   **185**

Literature   **195**

Danksagung   **209**

List of Abbreviations

ACPF	averaged coupled-pair functional (method)
AE	all-electron
AMFI	atomic mean-field integral (code)
An	actinides
ANO	atomic natural orbital
a.u.	atomic unit
avg.	average
BSSE	basis set superposition error
CASSCF	complete active space self-consistent field (method)
CASPT2	complete active space second-order perturbation theory
CC	coupled cluster (method)
CCSD(T)	coupled cluster method with singles and doubles and a perturbative estimate of triples
CI	configuration interaction
CISD+Q	configuration interaction including single and double excitations and the correction formula proposed by Langhoff and Davidson
COSMO	Conductor-like Screening Model
CP	counterpoise (correction)
CPP	core-polarization potential
CSF	configuration state function
CT	charge-transfer (excitation)
D	doubles (double excitations)
DaC	Davidson correction
DC	Dirac–Coulomb (Hamiltonian)
DCB	Dirac–Coulomb–Breit (Hamiltonian)
DFT	density functional theory

DHF	Dirac–Hartree–Fock (method)
DKH	Douglas–Kroll–Hess (Hamiltonian)
EA	electron affinity
ECP	effective core potential
EXAFS	extended X-ray absorption fine structure (spectroscopy)
exp.	experimental, experiment
FSCC	Fock-space coupled cluster (method)
GTO	Gaussian-type-orbital
Hba	benzoic acid C_6H_5COOH
Hbha	benzohydroxamic acid $C_6H_5CONHOH$
Hsha	salicylhydroxamic acid $HOC_6H_4CONHOH$
HF	Hartree–Fock (method)
HFS	Hartree–Fock–Slater (method)
hs	high-spin
IH-FSCC	intermediate Hamiltonian Fock-space coupled cluster (method)
IP	ionization potential
Ln	lanthanides
LPP	large-core pseudopotential
ls	low-spin
m.a.d.	mean absolute deviation
m.a.e.	mean absolute error
MCDF-CI	multi-configuration Dirac–Fock configuration interaction
MCDHF	multi-configuration Dirac–Hartree–Fock (method)
MCSCF	multi-configuration self-consistent field (method)
MO	molecular orbital
MO-LCAO	molecular orbitals by linear combination of atomic orbitals
MRCI	multi-reference configuration interaction
m.r.e.	mean relative error
MP	model potential
MP2	second-order Møller–Plesset perturbation theory
PES	potential energy surface
PP	pseudopotential
S	singles (single excitations)
SCF	self-consistent field (method)

SO	spin–orbit
SO-CI	spin–orbit configuration interaction (method)
SOMO	singly occupied molecular orbital
SPP	small-core pseudopotential
T	triples (triple excitations)
TD-DFT	time-dependent density functional theory
VE	valence electron
WB	Hartree–Fock method with scalar-relativistic corrections after Wood and Boring
XIH-FSCC	extrapolated intermediate Hamiltonian Fock-space coupled cluster (method)
ZORA	zero-order regular approximation

Nomenclature of Basis Sets:

VnZ, n=D, T, Q, ...	valence n-zeta basis set with n=double, triple, quadruple,...
aug	augmented
cc	correlation-consistent
p	polarized

For example:
aug-cc-pVQZ   augmented correlation-consistent valence quadruple-zeta basis set

Chapter 1

Introduction

The f-elements are divided into lanthanides Ln (La–Lu) and actinides An (Ac–Lr), for which the 4f and 5f shells are gradually filled with increasing nuclear charge, respectively. Lanthanides are also called rare earth elements, although they are quite abundant, i.e. even the least common lanthanide thulium (except for the radioactive praseodymium) is more abundant in the earth's crust than iodine [1]. They have many applications, e.g. cerium is used in lighter flints and many lanthanides are neutron absorbers in nuclear reactors [2]. Actinides, however, are very scarce and man-made except for Ac–U [1]. They are all toxic as well as radioactive and their applications are primarily power generation, nuclear weapons, and radiotherapy [3]. The main subjects of actinide chemistry are the improvement of the nuclear energy generation as well as the nuclear waste management, i.e. the selection of safe repositories and the reduction of the long-term radiotoxicity.

One possibility to reduce the long-term radiotoxicity of nuclear waste is the partitioning and transmutation strategy. Spent nuclear fuel contains the major actinides U and Pu, the minor actinides Np, Am, and Cm, and fission products, which are mainly lanthanides as Pm, Sm, and Eu [4]. While plutonium and the minor actinides are the smallest part of the spent fuel, they contribute most to the long-term radiotoxicity, i.e. spent fuel without reprocessing needs about one million years until its radiotoxicity decreases to that of the initial uranium, the separation of plutonium reduces this time to ca. 15000 years, and by additional removal of the minor actinides the radiotoxicity already can reach this value after less than 1000 years [4]. After the partitioning plutonium and the minor actinides can be transmuted to shorter-lived and/or less toxic species [4]. One of the key problems faced in partitioning and transmutation strategies is the separation of actinides and lanthanides, whose properties are very similar. Consequently a detailed understanding of actinide and lanthanide properties is highly desirable. However, these studies involve several difficulties for both experimental and theoretical work. While the toxicity, radioactivity, and scarcity of the actinides are the main obstacles for the experimentalists [3], theoreticians face particular challenges in the significant contributions of relativity as well as electron correlation for both lanthanides and actinides [5–8]. Furthermore, the complexities arising from partially occupied f shells make first-principle studies of lanthanide and actinide systems frequently cumbersome and motivate the development of approximate schemes based on chemical intuition [7, 8]. The number of theoretical papers between 1978 and 1992 for atoms and molecules containing f-elements documents these difficulties, i.e. for lanthanides less than 100 papers, and for heavier actinides as Am–Lr even less than ten papers were published in this period [9]. One exception is uranium, where slightly more than 100 publications appeared between 1978 and 1992 [9].

A commonly used approximation to cope with some of these problems in quantum chemical calculations is the effective core potential (ECP) approach, in which the explicit calculations are restricted to the chemically relevant valence electron system and relativistic effects are only implicitly accounted for by a proper adjustment of free parameters in the valence-only model Hamiltonian. In many cases only the application of ECPs allows for quantum mechanical calculations due to the computational savings. However, this method has to be seen as a suitable compromise between the computational effort and the accuracy of the results.

Essentially one distinguishes between two types of ECPs, i.e. model potentials (MPs) preserving the radial nodal structure of the all-electron (AE) valence orbitals and PPs, which use pseudo-valence-orbitals with a simplified radial nodal structure [10]. Today the most widely used variant of the MP method are the ab initio MPs of Huzinaga, Seijo, Barandiarán, and coworkers [11]. In the case of PPs one may further distinguish between shape-consistent and energy-consistent PPs, whereby the former are adjusted to orbital data of one reference configuration and the latter rely on the AE total valence energies of all chemical important configurations of the neutral atom and its low-charged ions [7]. A very popular set of shape-consistent PPs based on scalar-relativistic AE calculations was published by Hay and Wadt [12, 13]. Since in this thesis energy-consistent PPs of the Stuttgart-Cologne type will be presented, these PPs are discussed in more detail.

A fundamental decision prior to the construction of ECPs/PPs is the choice of the core and valence subsystems [7]. For f elements at least two kinds of energy-consistent PPs with different core definitions, i.e. 4f-in-valence [14] as well as 5f-in-valence [15, 16] small-core PPs (SPPs) and 4f-in-core [17, 18] as well as 5f-in-core [19, 20] large-core PPs (LPPs) are available. The f-in-valence SPPs treat 29–43 electrons explicitly, while 28 (1s–3d) and 60 (1s–4f) electrons are included in the PP core for lanthanides and actinides, respectively. The quasirelativistic Wood–Boring (WB) SPPs were already published almost 20 years ago [14, 15], and additionally a uranium SPP was recently adjusted to more rigorous AE four-component relativistic reference data [16]. Although the f-in-valence SPPs significantly reduce the computational effort, they do not avoid the difficulties due to the open f shell, which can lead to a spin and angular momentum as large as 7/2 and 12, respectively, resulting in many low-lying LS-states [21]. Due to these LS-states, which cause convergence problems, the calculation of large f-element complexes as lanthanide(III) texaphyrins [22] and actinide(III) motexafins [23] (cf. Fig. 1.1) were not feasible using f-in-valence SPPs. The 4f-in-core (1s–4f core) and 5f-in-core (1s–5f core) LPPs avoid many difficulties due to the open f shell, and despite their approximate nature are an efficient computational tool for those lanthanide/actinide compounds, where the f shell does not significantly contribute to chemical bonding. Using these LPPs the lanthanide(III) texaphyrin [22] and actinide(III) motexafin [23] complexes could be studied successfully.

While 4f-in-core LPPs for lanthanides were already adjusted in 1989 [17, 18], the first 5f-in-core LPPs for actinides were generated only some years ago [19, 20]. Both types of LPPs were adjusted at the quasirelativistic WB level, since a more accurate treatment of the relativistic makes no sense due to the crude approximation with respect to the core–valence separation. The reason for the delayed adjustment of the 5f-in-core LPPs can be explained by the oxidation states adopted by lanthanides and actinides in their compounds (cf. Table 1.1) [1, 2]. The preferred oxidation state in aqueous solution is +3 for all lanthanides, but only for the higher actinides (Am–Lr). For the early actinides the 5f shell can easily contribute to chemical bonding due to its diffuse character, and thus these actinides may reach formal oxidation states up to +7. The 4f shell has a core-like character for all lanthanides, while the 5f shell only becomes more core-like with increasing nuclear charge along the actinide series. Therefore the range of possible applications of the recently published 5f-in-core LPPs is certainly somewhat smaller than that for the 4f-in-core LPPs, which have already successfully been used during the last two decades by many researchers [8]. However, ample quantitative evidence is found that the 5f-in-core approximation can be made without too much loss of accuracy for many cases, e.g. actinide trifluorides [19, 20], actinide(III) mono- [19, 20] and polyhydrates [24], actinide(III) motexafin complexes [23], and crystalline uranium nitride [25].

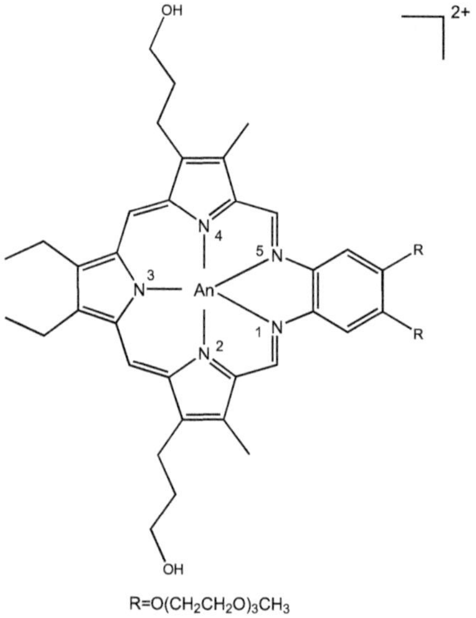

[Figure: Actinide(III) motexafin structure with R=O(CH$_2$CH$_2$O)$_3$CH$_3$, charge 2+]

R=O(CH$_2$CH$_2$O)$_3$CH$_3$

Figure 1.1: Actinide(III) motexafin structure.

In the case of the f-in-core LPPs one PP for each oxidation state, or rather, for each corresponding f subconfiguration is needed, since the frozen-core error shows a noticeable dependence on the f occupation (cf. Sect. 2.1.2) [7]. For lanthanides di- and trivalent 4f-in-core LPPs [17, 18] for La–Yb and La–Lu are available, respectively, and for actinides trivalent 5f-in-core LPPs [19, 20] for Ac–Lr were adjusted.

In this thesis the already existing f-in-core LPPs for actinides and lanthanides will be completed, i.e. di-[1], tetra-, penta-, and hexavalent 5f-in-core actinide LPPs for Pu–No, Th–Cf, Pa–Am, and U–Am, respectively, and tetravalent 4f-in-core lanthanide LPPs for Ce–Nd, Tb, and Dy will be adjusted. Corresponding basis sets for use in both

---

[1] The parameters of the divalent 5f-in-core PPs I have already adjusted during my diploma thesis [19].

Table 1.1: The oxidation states adopted by lanthanides and actinides in their compounds [1, 2]. The most stable oxidation state in aqueous solution is given in bold, and oxidation states only found in solids are given in parentheses.

La	Ce	Pr	Nd	Pm	Sm	Eu	Gd	Tb	Dy	Ho	Er	Tm	Yb	Lu
					(2)	2						(2)	(2)	
3	**3**	**3**	**3**	**3**	**3**	**3**	**3**	**3**	**3**	**3**	**3**	**3**	**3**	**3**
	4	(4)	(4)					(4)	(4)					

Ac	Th	Pa	U	Np	Pu	Am	Cm	Bk	Cf	Es	Fm	Md	No	Lr
						(2)			(2)	(2)	2	2	**2**	
3	3	(3)	3	3	3	**3**	**3**	**3**	**3**	**3**	**3**	**3**	3	**3**
	4	4	4	4	4	4	4	**4**	(4)					
		5	5	5	5	5								
			6	6	**6**	6								
				7	7									

crystal and molecular calculations will be optimized, and in the case of the actinides core-polarization potentials (CPPs) will be generated, in order to correct for the neglect of static and dynamic core-polarization. These PPs, CPPs, and basis sets will be tested in atomic and molecular calculations, and selected applications of the 5f-in-core LPPs, e.g. on actinocenes, actinyl ions, and uranyl(VI) complexes, will be presented. Furthermore, the recently adjusted 5f-in-valence uranium SPP, which includes scalar-relativistic effects as well as spin–orbit (SO) coupling, will be used to calculate the fine-structure splittings of U^{5+} and U^{4+}, in order to assess its accuracy.

Chapter 2

Theory

2.1 Effective Core Potentials

In the following first the motivation for ECPs and the main approximations of this method will be described. Afterwards different kinds of ECPs will be discussed, i.e. differences within the valence orbitals, the relativistic treatment, or the adjustment. Furthermore, the valence-only model Hamiltonian and the analytical form of ECPs will be given.

2.1.1 Motivation

The main reasons for the usage of ECPs are the computational savings resulting from the chemically intuitive restriction of the explicit calculations to the valence electron system and the implicit inclusion of the most important relativistic effects by means of a simple parameterization [7, 10, 26]. Since f-elements have many electrons and show large relativistic effects, calculations on molecules containing lanthanides or actinides are often only feasible, if ECPs are applied. Certainly the ECP approach requires some approximations (cf. Sect. 2.1.2), but it is a suitable compromise between the computational effort and the accuracy of the results.

If spin-dependent terms are averaged or neglected, the scalar-relativistic ECP calculations can be performed using non-relativistic quantum chemistry with only slight modifications [7, 26]. Even the transferability of ECPs (adjusted with wavefunction-based methods) to density functional theory (DFT) appears to be quite good [27], although the non-linearity of the total energy as well as the potential in the density is neglected [26].

Not necessarily every AE calculation yields a superior result. The reason for this is that ECPs allow to concentrate computational resources on the important parts of the system, i.e. they shift computational effort from the chemically unimportant core region to the valence electrons, which primarily determine the chemical behavior [10]. Thus, a higher quality treatment of the valence electron subsystem compared to the AE case becomes possible, e.g., larger basis sets and more accurate correlation methods can be applied [26]. Furthermore, the introduction of ECPs reduces the basis set superposition error (BSSE).

Finally, the ECP approach can help to avoid difficulties posed by open shells, if these are included in the core system. For lanthanides and actinides 4f-in-core [17, 18, 28] and 5f-in-core [20, 29, 30] PPs allow for calculations on large complexes as lanthanide(III) texaphyrins [22] and actinide(III) motexafins [23] (cf. Fig 1.1), respectively, which were not feasible due to convergence problems connected with the large amount of low-lying LS-states. However, these PPs can only be applied, if the f orbitals do not participate significantly in chemical bonding.

2.1.2 Approximations

A fundamental decision prior to the construction of ECPs is the choice of the core and valence subsystems or the so-called core–valence separation [7]. From a quantum mechanical point of view the partitioning of a many-electron system into subsystems is not possible, because electrons are indistinguishable [26]. However, in the framework of effective one-particle approximations as Hartree–Fock (HF) or Dirac–Hartree–Fock (DHF) theory such a core–valence separation is possible [26]. The definition of core and valence orbitals is either based on energetic or spatial arguments, i.e. on orbital energies or radial maxima of orbitals [26].

The core–valence separation involves at least two shortcomings. On the one hand the core-correlation consisting of core–core as well as core–valence correlation is neglected, i.e. if ECPs are applied, correlation is only accounted for the explicitly treated valence electrons, and even the static polarization of the core at the HF level is not considered. Due to the neglect of core-correlation bond lengths, ionization potentials (IPs), and bond energies are affected [31]. Obtained bond lengths and energies are too long and too small, respectively, because the electron–electron repulsion is overestimated, and calculated IPs are too small, since core-correlation stabilizes the atom.

2.1 EFFECTIVE CORE POTENTIALS

On the other hand the core orbitals are assumed to be transferable for the atom and molecules regardless of the electronic state, which corresponds to the frozen-core approximation [7, 26]. Therefore care has to be taken that all low-energy configurations of the neutral atom and low-charged ions, which might become important in chemical processes, are taken into account in the ECP adjustment [7,26]. However, only energy-consistent PPs can be adjusted to more than one configuration, which constitutes an advantage over shape-consistent PPs (cf. Sect. 2.1.6).

These two approximations are the more pronounced, the larger the ECP core. Thus, an appropriate choice of the ECP core is crucial. The validity of the frozen-core approximation can be assessed using AE calculations. Relativistic multi-configuration DHF (MCDHF) finite-difference calculations based on the Dirac–Coulomb (DC) Hamiltonian were performed for Th for a full variational solution as well as for three types of core definitions (cf. Table 2.1) [7]. From the chemical point of view the valence electron system of Th consists of four electrons $6d^2 7s^2$, which corresponds to a core–valence separation according to orbital energies. However, errors of more than 1 eV in ionization and excitation energies arise, especially if the 5f occupation number changes. This is due to the fact that the 6s and 6p semi-core orbitals are more diffuse than the compact 5f shell. Therefore the change of the 5f occupation will lead to a significant change of the effective nuclear charge for these orbitals and to a subsequent relaxation, which is not possible, if these orbitals are included in the ECP core. Furthermore, a weak dependence on the 6d occupation is observed, because this orbital still has a noticeable radial overlap with the 6s and 6p semi-core orbitals. A much better choice is to include the 6s and 6p semi-core orbitals in the valence space leading to ECPs with 12 valence electrons. In this case the dependence of the frozen-core errors on the 6d occupation is negligible (at most 0.005 eV), whereas a noticeable systematic dependence of at most 0.272 eV on the 5f occupation is still present. These findings can be explained by the radial overlap between the 5f valence and 5s, 5p, and 5d core shells. The best choice with respect to the frozen-core error is a small-core approach, which separates core and valence space on the basis of spatial arguments treating all electrons with main quantum number larger than four explicitly, i.e. 30 valence electrons. The frozen-core errors are at most 0.001 eV. However, the accuracy of these small-core potentials is traded against the low computational cost of the large-core potentials (12 valence electrons).

Table 2.1: Relative DHF energies (in eV) of the 2J+1-weighted average of all J levels belonging to a non-relativistic configuration with respect to the value for the Th [Rn]$6d^2 7s^2$ ground state configuration (Table 3 from [7]). Only subconfigurations outside the Rn core are listed. The frozen-core errors (in eV) in the relative energies are given for 4, 12, and 30 valence electron (VE) systems. The frozen-core was taken from the neutral atom in the ground state configuration.

					Frozen-core Error		
	Configuration			DHF	4VE	12VE	30VE
				60.301	1.113	0.005	0.000
			$7s^1$	35.394	0.593	0.002	0.000
			$7s^2$	17.357	0.292	0.002	0.000
		$6d^1$		34.179	0.382	0.001	0.000
		$6d^1$	$7s^1$	16.505	0.154	0.000	0.000
		$6d^1$	$7s^2$	5.151	0.051	0.000	0.000
		$6d^2$		16.516	0.075	0.001	0.000
		$6d^2$	$7s^1$	5.434	0.013	0.000	0.000
		$6d^2$	$7s^2$	0.000	0.000	0.000	0.000
		$6d^3$		6.503	0.022	0.001	0.000
		$6d^3$	$7s^1$	1.206	0.015	0.001	0.000
		$6d^4$		3.055	0.051	0.002	0.000
$5f^1$				33.873	0.235	0.071	0.000
$5f^1$			$7s^1$	16.860	0.206	0.080	0.000
$5f^1$			$7s^2$	6.043	0.233	0.085	0.000
$5f^1$		$6d^1$		17.279	0.292	0.065	0.000
$5f^1$		$6d^1$	$7s^1$	6.715	0.328	0.070	0.000
$5f^1$		$6d^1$	$7s^2$	1.645	0.358	0.072	0.000
$5f^1$		$6d^2$		8.113	0.443	0.056	0.000
$5f^1$		$6d^2$	$7s^1$	3.138	0.442	0.059	0.000
$5f^1$		$6d^3$		5.186	0.514	0.048	0.000
$5f^2$				20.073	0.997	0.226	0.001
$5f^2$			$7s^1$	10.028	1.022	0.221	0.001
$5f^2$			$7s^2$	5.327	0.961	0.208	0.001
$5f^2$		$6d^1$		11.628	1.084	0.179	0.001
$5f^2$		$6d^1$	$7s^1$	6.952	0.862	0.164	0.001
$5f^2$		$6d^2$		9.010	0.606	0.112	0.001
$5f^3$				16.336	1.375	0.272	0.001
$5f^3$			$7s^1$	11.444	0.850	0.195	0.001
$5f^3$		$6d^1$		13.081	0.698	0.128	0.001

2.1 Effective Core Potentials

A possible correction for the neglect of static and dynamic core-polarization arising from the core–valence separation and the frozen-core approximation is the introduction of CPPs [26], which are especially important for systems with easily polarizable cores as LPPs. Classically, the core–valence effect is attributed to the polarization of the core electron system by the valence electrons and other cores. Müller et al. [32,33] proposed in the framework of AE calculations the use of an effective CPP of the form [7, 26]

$$V_{CPP} = -\frac{1}{2}\sum_I \alpha_D^I \vec{f}_I^2 \qquad (2.1)$$

$$\text{with} \quad \vec{f}_I = -\sum_i \frac{\vec{r}_{iI}}{r_{iI}^3}\omega(r_{iI}) + \sum_{J\neq I} Q_J \frac{\vec{R}_{JI}}{R_{JI}^3}\omega(R_{JI}) \qquad (2.2)$$

$$\text{and} \quad \omega(r) = \left(1 - \exp\left(-\delta r^2\right)\right)^n. \qquad (2.3)$$

Here α_D^I denotes the dipole polarizability of the core I and $\vec{f}_I$ is the electric field at this core generated by the valence electrons (at relative positions $\vec{r}_{iI}$) and all other cores or nuclei (with charges Q_J, at relative positions $\vec{R}_{JI}$). Since the validity of the underlying multipole expansion breaks down for small distances from the core I, the electric field $\vec{f}_I$ has to be multiplied by a cutoff factor ω. This ansatz was adapted by the Stuttgart group [34–36] for the PP case and proved to be quite successful in calculations using energy-consistent LPPs of main group elements [37, 38], group 11 and 12 transition metals [39, 40], and lanthanides [41].

Another approximation is the replacement of the core electron system via the introduction of an ECP, i.e. the usage of the valence-only model instead of the AE Hamiltonian [9]. If, e.g., the Dirac one-particle Hamiltonian (2.9) is substituted by the corresponding non-relativistic expression (2.7), relativistic effects are only accounted for the core electron system by means of the ECP, while valence electrons are treated non-relativistically. However, it is a widely accepted fact that for not too highly-charged cores the valence electron system can be treated in a formally non-relativistic scheme [9].

2.1.3 Model Potential vs. Pseudopotential

One distinguishes two main lines of ECP approaches, i.e. the MP and the PP techniques [7, 26]. The MP approach uses valence orbitals with a nodal structure corresponding exactly to those of the AE valence orbitals and shifts the (now unoccupied) core-like orbitals to the virtual space. As an additional approximation the PP scheme (formally) introduces the so-called pseudo-valence-orbital transformation, i.e. atomic core and virtual orbitals are mixed into the valence orbitals, in order to make these radially smooth and nodeless for the energetically lowest solution in each angular symmetry. Although the pseudo-valence-orbitals possess in the chemically inert core region a simplified nodal structure compared to the AE or MP valence orbitals, their shapes in the chemically important valence region as well as their one-particle energies should be similar to the AE case [7].

Experience from several benchmark studies shows that both approaches are able to yield results in excellent agreement with more rigorous AE methods. However, both schemes have advantages and disadvantages. The exact nodal structure of the MP valence orbitals in the core region requires compact basis functions, which are usually not needed in the description of chemical bonding. Therefore considerable savings with respect to the one-particle basis set can be achieved, if the explicit requirement of core–valence orthogonality is given up, introducing pseudo-valence-orbitals with simplified nodal structure by including the necessary corrections into the valence-only model Hamiltonian. However, pseudo-valence-orbitals tend to give too large valence correlation energies as well as too large multiplet splittings, since the exchange integrals involving these orbitals are overestimated. In practice the accuracy of correlation energies from PP calculations is not worse than that from MP calculations and especially correlation contributions to energy differences as binding energies are well reproduced due to the modern PP parameterization [26].

2.1.4 Relativistic Treatment

In general the task of quantum chemistry is to solve the time-dependent Schrödinger equation

$$\hat{H}\Psi = i\hbar \frac{\partial}{\partial t}\Psi, \qquad (2.4)$$

where $\Psi(x, y, z, t)$ is the wave function, which contains all informations about a system. Fortunately, for many applications of quantum mechanics to chemistry the calculation of stationary states $\psi(x, y, z)$ is sufficient, i.e. the solution of the simpler time-independent Schrödinger equation

$$\hat{H}\psi = E\psi \qquad (2.5)$$

2.1 EFFECTIVE CORE POTENTIALS

is needed, in order to determine the energy eigenvalue E [42]. Supposing that the movement of the nuclei are negligible compared to that of the electrons (Born–Oppenheimer approximation) and that there are no external fields, the Hamiltonian[1] $\hat{H}$ for a system consisting of n electrons with indices i, j and N nuclei with charges Q and indices I, J is given by

$$\hat{H} = \sum_{i=1}^{n} \hat{h}(i) + \sum_{i<j}^{n} \hat{g}(i,j) + \sum_{I<J}^{N} \frac{Q_I Q_J}{R_{IJ}} . \qquad (2.6)$$

For the one- $\hat{h}$ and two-particle $\hat{g}$ operators various expressions can be inserted, e.g. non-relativistic, quasirelativistic, or relativistic as well as AE or valence-only formulations [26]. The last term of (2.6) corresponds to the nucleus–nucleus repulsion with R_{IJ} being the distance between (point-charge) nuclei I and J. This potential energy is independent of the electronic wavefunction, i.e. it is a constant for a given nuclear configuration. Thus, the internuclear repulsion can be omitted from (2.5), and after finding the energy eigenvalue, it can be added to yield the total energy [42].

How far ECPs include relativistic effects depends on the Hamiltonian chosen to calculate the AE reference data. For the non-relativistic Hamiltonian [43] the one-particle operator $\hat{h}$ is the Schrödinger Hamiltonian

$$\hat{h}_S(i) = -\frac{1}{2}\Delta_i - \sum_{I=1}^{N} \frac{Q_I}{r_{iI}} \qquad (2.7)$$

and the two-particle operator $\hat{g}$ corresponds to the electrostatic Coulomb interaction between electrons

$$\hat{g}_C(i,j) = \frac{1}{r_{ij}} . \qquad (2.8)$$

The first term of $\hat{h}_S$ describes the kinetic energy of the electron, while the second one corresponds to the electron–nucleus attraction with r_{iI} being the distance between electron i and (point-charge) nucleus I. This non-relativistic Hamiltonian is the basis of the HF method (cf. Sect. 2.2.1) and yields non-relativistic ECPs, if used to calculate the reference data.

---
[1] All formulas are given in atomic units (a.u.).

The most accurate Hamiltonian nowadays is based on the Dirac one-particle Hamiltonian [7]

$$\hat{h}_D(i) = c\hat{\vec{\alpha}}_i\hat{\vec{p}}_i + (\hat{\beta} - I_4)c^2 - \sum_{I=1}^{N} \frac{Q_I}{r_{iI}}, \qquad (2.9)$$

where the rest energy of the electron c^2 was subtracted, in order to have the same zero of energy as in the non-relativistic case. In this equation c is the light velocity, $\hat{\vec{p}}_i = -i\vec{\nabla}_i$ corresponds to the momentum operator for the i-th electron, and I_4 denotes the 4×4 unit matrix. $\hat{\vec{\alpha}}_i$ is a three-component vector, whose elements together with $\hat{\beta}$ are the 4×4 Dirac matrices

$$\hat{\vec{\alpha}} = \begin{pmatrix} 0 & \hat{\vec{\sigma}} \\ \hat{\vec{\sigma}} & 0 \end{pmatrix} \quad \text{and} \quad \hat{\beta} = \begin{pmatrix} I_2 & 0 \\ 0 & -I_2 \end{pmatrix}. \qquad (2.10)$$

These can be expressed in terms of the three-component vector of the 2×2 Pauli matrices $\hat{\vec{\sigma}}$

$$\hat{\sigma}_x = \begin{pmatrix} 0 & 1 \\ 1 & 0 \end{pmatrix}, \quad \hat{\sigma}_y = \begin{pmatrix} 0 & -i \\ i & 0 \end{pmatrix}, \quad \hat{\sigma}_z = \begin{pmatrix} 1 & 0 \\ 0 & -1 \end{pmatrix}, \qquad (2.11)$$

and the 2×2 unit matrix I_2. In contrast to the Dirac one-particle Hamiltonian (2.9), which is exact to all orders of the fine-structure constant $\alpha = 1/c$, only approximate expressions are known for the two-particle term $\hat{g}$ [7]. If this term is chosen to be the non-relativistic Coulomb interaction (2.8), the DC Hamiltonian $\hat{H}_{DC}$ correct to the zero-order of the fine-structure constant

$$\hat{H}_{DC} = \hat{h}_D(i) + \hat{g}_C(i,j) \qquad (2.12)$$

is obtained. The DHF as well as the MCDHF method are based on this Hamiltonian, which can be used to adjust relativistic ECPs. In addition the magnetic interaction and the retardation of the interaction due to the finite light velocity can be included using the frequency-independent Breit interaction [7]

$$\hat{g}_{CB}(i,j) = \frac{1}{r_{ij}} - \frac{\hat{\vec{\alpha}}_i \hat{\vec{\alpha}}_j}{2r_{ij}} - \frac{(\hat{\vec{\alpha}}_i \vec{r}_{ij})(\hat{\vec{\alpha}}_j \vec{r}_{ij})}{2r_{ij}^3}. \qquad (2.13)$$

2.1 EFFECTIVE CORE POTENTIALS

The Dirac–Coulomb–Breit (DCB) Hamiltonian

$$\hat{H}_{DCB} = \hat{h}_D(i) + \hat{g}_{CB}(i,j) \qquad (2.14)$$

is correct to the second-order of the fine-structure constant. The contributions of higher-order corrections, e.g. the vacuum polarization or self-energy of the electron, can be derived from quantum electrodynamics, but are usually neglected. Since the Breit interaction is small compared to the relativistic effects of the Dirac one-particle operator (2.9), often it is not treated variationally, but rather perturbatively after a variational treatment of the DC Hamiltonian [7]. The 5f-in-valence uranium SPP [16] tested in this work was adjusted to such four-component AE MCDHF/DCB data [44] using a Fermi two parameter nucleus charge distribution $\rho(r)$ [44, 45]

$$\rho(r) = [1 + \exp((r - R_{nuc})/t)]^{-1} . \qquad (2.15)$$

The nuclear radius is derived from the nuclear mass M by $R_{nuc} = 0.836 \cdot 10^{-15}$ m $\cdot M^{1/3} + 0.570 \cdot 10^{-15}$ m, and the skin depth is $t = 2.30 \cdot 10^{-15}$ m.

However, the four-component methods face several difficulties due to the Dirac one-particle Hamiltonian (2.9), because it is not bounded from below and gives rise to so-called electronic and positronic states [7]. Therefore an energy-variation without additional precautions could lead to a variational collapse of the desired electronic solution into the lower-energy positronic states [26]. Additionally, at the many-electron level an infinite number of unbound states with one electron in the positive and one in the negative continuum are degenerate with the desired solution having all electrons in bound electronic states [7]. A mixing-in of these unphysical states is possible without changing the energy and might lead to the so-called continuum dissolution or Brown–Ravenhall disease [26]. Both problems are avoided by projecting the Hamiltonian onto the electronic states by means of suitable operators $\hat{P}_+$. So the no-pair Hamiltonian

$$\hat{H}_{np} = \hat{P}_+ \hat{H} \hat{P}_+ \qquad (2.16)$$

is obtained, which is approximate due to the underlying DCB Hamiltonian as well as the approximate nature of the DHF or MCDHF based projection operators $\hat{P}_+$ [7]. Because of these difficulties PPs are often adjusted to quasirelativistic reference data, e.g., obtained using the WB method [46], which works within the LS coupling scheme.

The 5f-in-core [20,29,30] and 4f-in-core [17,18,28] LPPs adjusted and/or investigated in this thesis are one-component quasirelativistic PPs based on AE WB reference data. Within the central field approximation, where the electron–nucleus attraction $V(r)$ is spherically symmetric, for a one-electron atom one obtains the radial WB equation by elimination of the small components from the Dirac equation [26]

$$\left(\hat{h}_S + \hat{h}_{MV} + \hat{h}_D + \hat{h}_{SO}\right) P_{n\kappa}(r) = \epsilon_{n\kappa} P_{n\kappa}(r) . \qquad (2.17)$$

In this second-order differential equation for the large components $P_{n\kappa}(r)$ besides the non-relativistic Schrödinger Hamiltonian

$$\hat{h}_S = -\frac{1}{2}\frac{d^2}{dr^2} + \frac{l(l+1)}{2r^2} + V(r) \qquad (2.18)$$

three energy-dependent relativistic terms occur, i.e. a mass-velocity $\hat{h}_{MV}$, a Darwin $\hat{h}_D$, and a SO $\hat{h}_{SO}$ term

$$\hat{h}_{MV} = -\frac{\alpha^2}{2}[\epsilon_{n\kappa} - V(r)]^2 , \qquad \hat{h}_D = -\frac{\alpha^2}{4}\frac{dV}{dr}B_{n\kappa}\left(\frac{d}{dr} - \frac{1}{r}\right) , \qquad (2.19)$$

$$\hat{h}_{SO} = -\frac{\alpha^2}{4}\frac{dV}{dr}B_{n\kappa}\frac{\kappa+1}{r} , \qquad B_{n\kappa} = \left(1 + \frac{\alpha^2}{2}[\epsilon_{n\kappa} - V(r)]\right)^{-1} .$$

Equation (2.17) can be solved iteratively and yields the exact (electronic) eigenvalues $\epsilon_{n\kappa}$ of the corresponding Dirac equation. Averaging over the relativistic quantum number κ

$$\kappa = \begin{cases} -(l+1) & \text{for } j = l+1/2 \\ l & \text{for } j = l-1/2 \end{cases} \qquad (2.20)$$

leads to a scalar-relativistic scheme.

2.1.5 Valence-only Model Hamiltonian

In ECP theory an effective model Hamiltonian approximation for the no-pair Hamiltonian $\hat{H}_{np}$ (2.16) is searched, which only acts on the electronic states formed by the valence electrons. Several choices exist for such a valence-only model Hamiltonian, i.e. four-, two-, or one-component approaches and explicit or implicit relativistic treatment. Since a reasonable compromise between accuracy and efficiency is desired, normally the one-component (scalar-relativistic) treatment and the implicit relativistic

2.1 Effective Core Potentials

treatment by ECPs using a non-relativistic Hamiltonian are used. The formally non-relativistic valence-only model Hamiltonian for a system with n_ν valence electrons and N cores with effective core charges Q is given as [26]

$$\hat{H}_\nu = -\frac{1}{2}\sum_{i=1}^{n_\nu}\Delta_i + \sum_{I=1}^{N} V_{c\nu}^I(r_{iI}) + \sum_{i<j}^{n_\nu}\frac{1}{r_{ij}} + \sum_{I<J}^{N}\frac{Q_I Q_J}{R_{IJ}} + V_{CPP} \,. \tag{2.21}$$

Here, the subscripts ν and c denote valence and core, respectively, and V_{CPP} is a CPP (2.1). The number of valence electrons treated explicitly in the calculations corresponds to

$$n_\nu = n - \sum_{I}^{N}(Z_I - Q_I)\,, \tag{2.22}$$

where Z_I and Q_I denote the atomic number and the effective core charge of core I, respectively. The scalar-relativistic one-component ECP $V_{c\nu}^I$ describes the interaction of the valence electrons with core I and its semilocal form is

$$V_{c\nu}^I(r_{iI}) = -\sum_{i=1}^{n_\nu}\frac{Q_I}{r_{iI}} + \sum_{i=1}^{n_\nu} V_I^{PP}(r_{iI}) = -\sum_{i=1}^{n_\nu}\frac{Q_I}{r_{iI}} + \sum_{i=1}^{n_\nu}\sum_{l=0}^{l_{max}}\sum_{k} A_{lk}^I \exp(-a_{lk}^I r_{iI}^2)\hat{P}_l^I \,. \tag{2.23}$$

The PPs V_I^{PP} are represented as linear combinations of k Gaussians for each angular quantum number l included in the core, and

$$\hat{P}_l^I = \sum_{m_l=-l}^{l} |Ilm_l\rangle\langle Ilm_l| \tag{2.24}$$

is the projection operator onto the Hilbert subspace of core I with angular quantum number l.

While the f-in-core PPs [17,18,20,28–30] are of this scalar-relativistic one-component type, the 5f-in-valence PP for uranium [16] is a quasirelativistic (including SO coupling) two-component (semilocal) PP of the form [10]

$$V_{c\nu}^I(r_{iI}) = \sum_{l=0}^{L-1}\sum_{j=|l-1/2|}^{l+1/2} V_{lj}^I(r_{iI})\hat{P}_{lj}^I \,. \tag{2.25}$$

At the non-relativistic HF level all orbitals belonging to a shell with main quantum number n and angular quantum number l are degenerate, thus leading to a PP depending on l by means of a projection operator $\hat{P}_l^I$ based on spherical harmonics. At the relativistic DHF level the degeneracy is reduced and depends additionally to n and l on the total angular momentum j of the orbital (or spinor), i.e. here the PP depends on l and j by means of a projection operator $\hat{P}^I_{lj}$ set up with spinor spherical harmonics $|Iljm_j\rangle$ [10]

$$\hat{P}^I_{lj} = \sum_{m_j=-j}^{j} |Iljm_j\rangle\langle Iljm_j| \,. \qquad (2.26)$$

The relativistic PP (2.25) may be written as the sum of a spin-free averaged and a spin-dependent SO-term [10]

$$V^I_{cv}(r_{iI}) = V^I_{cv,avg}(r_{iI}) + V^I_{cv,SO}(r_{iI})\,, \qquad (2.27)$$

where

$$V^I_{cv,avg}(r_{iI}) = \frac{(l+1)V^I_{l,l+1/2}(r_{iI}) + lV^I_{l,l-1/2}(r_{iI})}{2l+1} \qquad (2.28)$$

and

$$V^I_{cv,SO}(r_{iI}) = \sum_{l=1}^{L-1} \frac{V^I_{l,l+1/2}(r_{iI}) - V^I_{l,l-1/2}(r_{iI})}{2l+1} \left[l\hat{P}^I_{l,l+1/2} - (l+1)\hat{P}^I_{l,l-1/2} \right]. \qquad (2.29)$$

The semilocal terms V^I_{lj} of the two-component PP (2.25) are as usual represented as linear combinations of k Gaussians [47]

$$V^I_{lj}(r_{iI}) = \sum_{k} B^I_{ljk} \exp\left(-b^I_{ljk} r^2_{iI}\right) \,. \qquad (2.30)$$

2.1.6 Energy Adjustment

Among the PPs one may distinguish shape-consistent and energy-consistent PPs. While the former are adjusted to orbital data (orbital shape in the valence region and orbital energy) of one reference configuration, the latter rely on the AE total valence energies of all chemical important configurations of the neutral atom and its low-charged ions, which are quantum mechanical observables [7]. Thus, energy-consistent PPs better account for the transferability of the core orbitals, the so-called frozen-core approximation (cf. Sect. 2.1.2).

The energy adjustment consists of three steps [7]. First, the reference configurations I are chosen and their total energies E^{AE}_I are determined using an AE method. Next the AE total valence energies $E^{AE,V}_I$ are derived by subtracting the core energy from

2.1 EFFECTIVE CORE POTENTIALS

the total energies E_I^{AE}. Finally, the free parameters (coefficients and exponents of the Gaussians) of the PP are adjusted by a least-squares fit to the total valence energies of the reference states. In the case of the WB f-in-core LPPs the sum of weighted-squared errors in the total valence energies E_I^{PP} with respect to the unmodified (ΔE_{shift}=0) AE total valence energies $E_I^{AE,V}$ is minimized

$$\sum_I \left(\omega_I [E_I^{PP} - E_I^{AE,V} + \Delta E_{shift}]^2 \right) := \min . \qquad (2.31)$$

The weights ω_I are typically chosen to be equal for all reference configurations, and the requirements for the accuracy are 0.1 eV for the total valence energies of many-electron configurations with one or two Gaussians per radial potential of each l-value included in the core [10].

In contrast to this for the adjustment of the MCDHF/DCB 5f-in-valence uranium SPP a global valence energy shift ΔE_{shift} was introduced as an additional adjustable parameter (cf. Fig. 2.1) [10, 16, 47]. Whereas the restriction to ΔE_{shift}=0 implied that, e.g., the ground state valence energy equals the sum of all IPs leading from the neutral atom to the core-electron system, this is not the case for the new fitting procedure. Here only the sum of all IPs leading from the neutral atom to the most highly ionized system included in the adjustment is reproduced correctly. The shift typically amounts to 1% or less of the ground state total valence energy and can improve the accuracy of the adjustment by one or two orders of magnitude, e.g. for the adjustment of the U SPP with up to four Gaussians in each radial potential an accuracy of better than 0.01 eV for configurational averages was achieved (cf. Fig. 3.22) [16]. Furthermore, thanks to ΔE_{shift} the adjustment to higher ionized states even with holes in core and semi-core orbitals becomes possible.

The shift can also be viewed as a shift of the AE core energy as shown in Fig. 2.1 for the uranium SPP with 60 core and 32 valence electrons, respectively [10]. Since the bare core position relative to the valence states is not expected to be overly relevant for chemical processes, this shift changing the reference energies can be justified [48]. Moreover, it is obvious that the quantities of interest as the electron affinity, IPs, and excitation energies, i.e. all possible energy differences between configurations included as references remain unchanged [10].

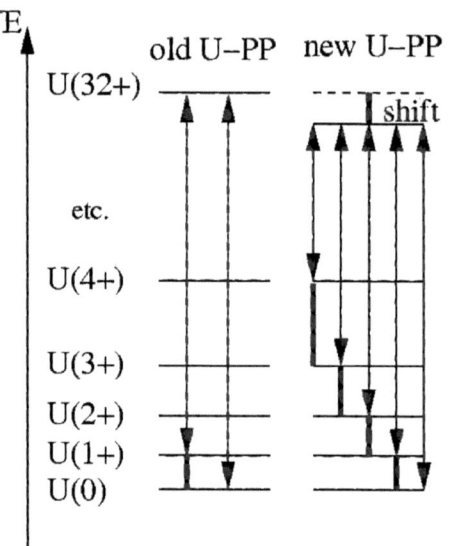

Figure 2.1: Schematic presentation of the reference data (arrows) as well as the data usually of interest (bars) in the energy adjustment of the U 5f-in-valence SPP. In contrast to the old WB for the new MCDHF/DCB SPP a shift of the core energy ΔE_{shift} was included in the adjustment (Fig. 1.2 from [10] modified for the U SPP).

2.2 Quantum Mechanical Methods

In general computational chemistry tries to solve the time-independent non-relativistic Schrödinger equation (2.5), in order to determine the energy eigenvalue and other physical properties of an atom or molecule. The four main approaches are ab initio methods, semiempirical methods, DFT, and molecular-mechanics [42]. Semiempirical methods as the Hückel molecular orbital method use a simplified Hamiltonian and parameters, which are adjusted to experimental data or the results of ab initio calculations. In contrast an ab initio wave function based calculation applies the correct Hamiltonian and does not use any experimental data. The DFT calculates the electronic energy using the electron probability density instead of the wave function. Molecular-mechanics, however, are not quantum mechanical methods, because they do not deal with a Hamiltonian, wave function, or electron density, but view molecules as a collection of atoms held together by bonds and express the energy in terms of force constants for bond bending and stretching and other parameters. Because molecular-mechanics are much faster than quantum mechanical calculations, systems with up to ten thousand atoms can be treated, wherefore these methods are, e.g., applied to determine solvation effects.

For the ab initio wave function based methods the exact relation between the wave function ψ and the energy eigenvalue E is the well-known (non-relativistic) Hamiltonian (cf. (2.6)–(2.8)). However, the wave function $\psi(x_1, y_1, z_1, \sigma_1, ...)$ depends on $4n$ variables, i.e. the three spatial coordinates and the spin σ of the n electrons. Since the Hamiltonian contains only one- and two-electron spatial terms, the molecular energy can be calculated by integrals involving only six spatial coordinates [42]. Thus, the wave function comprises more information than needed. In 1964 Hohenberg and Kohn proved that for molecules the ground state energy, wave function, and all other electronic properties are uniquely determined by the ground state electron probability density $\rho_0(x, y, z)$, which is a function of three coordinates (Hohenberg–Kohn theorem) [49]. However, in the case of DFT the functional relation between the electronic energy E_0 and the electron probability density ρ_0 is not known and has to be approximated.

In the following the applied (non-relativistic) ab initio methods (the HF method, configuration interaction (CI), the multi-configuration self-consistent field (MCSCF) method, the second-order Møller–Plesset perturbation theory (MP2), and the coupled cluster (CC) theory) as well as the DFT will be described in detail. The Born–Oppenheimer approximation (cf. Sect. 2.1.4) is assumed to hold and thus only the solution of the electronic Schrödinger equation for a given nuclear configuration is shown.

2.2.1 Hartree–Fock Method

The Schrödinger equation for a many-electron system (cf. (2.5)–(2.8)) cannot be solved exactly due to the electron–electron repulsion (2.8), which couples the motions of the electrons and thus makes a separation of this Schrödinger equation impossible. However, supposed that each electron moves in the average field of the others

$$\sum_{i<j}^{n} \frac{1}{r_{ij}} \rightarrow \sum_{i=1}^{n} \hat{v}^{HF}(i) , \quad (2.32)$$

the problem can be decoupled. Due to this approximation the n-electron wave function $\psi(1, 2, ..., n)$ is factorized into n independent one-electron wave functions so-called (spin)orbitals ϕ_i. In the HF method the n-electron wave function is approximated by a Slater determinant

$$\Phi = \frac{1}{\sqrt{n!}} \sum_{k=1}^{n!} (-1)^{p_k} \hat{P}_k \prod_{i=1}^{n} \phi_i(i) , \quad (2.33)$$

which accounts for the fact that electrons are indistinguishable and which satisfies the Pauli principle, i.e. the requirement that the wave function must be antisymmetric with respect to interchange of any two electrons [42]. Here $\hat{P}_k$ is the permutation operator, which constitutes all $n!$ possible configurations of the n electrons in the n orthonormal spinorbitals ϕ_i. p_k accounts for the antisymmetry requirement, i.e. it corresponds to the number of interchanges needed to get the kth permutation. For n orthonormal spinorbitals ($\langle \phi_i | \phi_i \rangle = \delta_{ij}$) the normalization constant is $1/\sqrt{n!}$.

In order to determine the spinorbitals, the variation theorem is exploited, i.e. the fact that any (normalized) trial function $\tilde{\psi}_0$ yields an energy expectation value $\tilde{E}_0$, which is larger or in the best case ($\tilde{\psi}_0 = \psi_0$) equal to the true ground state energy E_0 calculated using the exact (normalized) ground state wave function ψ_0

$$E_0 = \langle \psi_0 | \hat{H} | \psi_0 \rangle \leq \langle \tilde{\psi}_0 | \hat{H} | \tilde{\psi}_0 \rangle = \tilde{E}_0 . \quad (2.34)$$

Thus, the variation theorem allows to calculate an upper bound to the true ground state energy of the system. The best trial function or Slater determinant is obtained by varying the spinorbitals ϕ_i to minimize $\tilde{E}_0$ ($\partial \tilde{E}_0 / \partial \phi_i = 0$).

2.2 QUANTUM MECHANICAL METHODS

Turning points of a function subject to constraints can be assigned using the method of Lagrange multipliers. For the minimum of $\tilde{E}_0$ with the constraint of orthonormal spinorbitals, one gets n one-electron equations, the so-called HF equations

$$\hat{f}\phi_i = \epsilon_i \phi_i \ . \tag{2.35}$$

Since the Fock operator

$$\hat{f} = -\frac{1}{2}\Delta_i - \sum_{I=1}^{N} \frac{Q_I}{r_{iI}} + \hat{v}^{HF}(i) = -\frac{1}{2}\Delta_i - \sum_{I=1}^{N} \frac{Q_I}{r_{iI}} + \sum_{j=1}^{n} \left[\hat{J}_j(i) - \hat{K}_j(i) \right] \tag{2.36}$$

depends on its solutions, i.e. the spinorbitals ϕ_i, the spinorbital energies ϵ_i can only be determined iteratively using a self-consistent field (SCF) procedure. The sum of the third term involves all occupied spinorbitals j, and the Coulomb J_{ij} and exchange integrals K_{ij} corresponding to the Coulomb $\hat{J}_j$ and exchange $\hat{K}_j$ operators are given by

$$J_{ij} = \langle \phi_i | \hat{J}_j | \phi_i \rangle = \int_{-\infty}^{\infty} \int_{-\infty}^{\infty} \phi_i^*(1)\phi_j^*(2)\frac{1}{r_{12}}\phi_i(1)\phi_j(2)d\tau_1 d\tau_2 \tag{2.37}$$

and

$$K_{ij} = \langle \phi_i | \hat{K}_j | \phi_i \rangle = \int_{-\infty}^{\infty} \int_{-\infty}^{\infty} \phi_i^*(1)\phi_j^*(2)\frac{1}{r_{12}}\phi_j(1)\phi_i(2)d\tau_1 d\tau_2 \ , \tag{2.38}$$

respectively. While the Coulomb operator accounts for the electrostatic Coulomb interaction between electrons, the exchange operator takes into account the effects of spin correlation [50], i.e. it includes the reduction of the Coulomb energy due to the fact that electrons with identical spins cannot occupy the same position. Since for $i=j$ Coulomb and exchange integrals are equal ($J_{ii}=K_{ii}$), the electron self-interaction is eliminated from $\hat{v}^{HF}$.

Whereas the HF equations for atoms can still be solved numerically, an additional approximation is needed to calculate molecular wave functions, i.e. the representation of spinorbitals ϕ_i as linear combinations of basis functions χ_k proposed by Roothaan in 1951 [42]

$$\phi_i = \sum_{k=1}^{k_{max}} c_{ki} \chi_k \ . \tag{2.39}$$

The expansion coefficients c_{ki} are determined by the SCF procedure. The more basis functions are utilized, the more accurate is the HF solution, which for an infinite number of basis functions ($k_{max}=\infty$) would correspond to the exact numerical solution, the HF limit. If k_{max} is large enough and the basis functions χ_k are well chosen, the spinorbitals are represented with negligible error [42].

The introduction of basis functions leads to the Roothaan equation

$$\underline{\underline{F}}\,\underline{\underline{C}} = \underline{\underline{S}}\,\underline{\underline{C}}\,\underline{\underline{\epsilon}}, \qquad (2.40)$$

which is a matrix equation, where $\underline{\underline{F}}$, $\underline{\underline{C}}$, and $\underline{\underline{S}}$ correspond to the Fock, coefficient, and overlap square matrices, respectively, and $\underline{\underline{\epsilon}}$ is a diagonal square matrix of the spinorbital energies ϵ_i. Since the basis functions χ_i are not orthogonal, the overlap matrix $\underline{\underline{S}}$ arises in (2.40), wherefore this equation is not a usual matrix eigenvalue problem. However, the basis functions can be orthogonalized using, e.g. the Schmidt procedure, i.e. the overlap matrix becomes a unit matrix and the simpler equation

$$\underline{\underline{\tilde{F}}}\,\underline{\underline{\tilde{C}}} = \underline{\underline{\tilde{C}}}\,\underline{\underline{\epsilon}} \qquad (2.41)$$

is obtained [42]. Thus, instead of the complicated integro-differential equations (2.35)–(2.38) only a matrix eigenvalue problem (2.41) has to be solved using the Roothaan–Hall formalism.

After iterative determination of $\underline{\underline{\epsilon}}$ and thus of the spinorbital energies ϵ_i, the total HF energy is calculated by

$$E_{HF} = \sum_{i=1}^{n} \epsilon_i - \sum_{i<j}^{n}(J_{ij} - K_{ij}) + \sum_{I<J}^{N} \frac{Q_I Q_J}{R_{IJ}}. \qquad (2.42)$$

Here the second term avoids that the electron–electron repulsion is counted twice, and the third term corresponds to the so far neglected nucleus–nucleus repulsion. The HF energy E_{HF} is typically by ca. 0.5% too large, wherefore quantitative conclusions, e.g. for binding energies, cannot be drawn [42]. The reason for this deviation is that only the mean interaction between electrons with different spins is taken into account, while the spin correlation, i.e. the deviations from a mean interaction between electrons with equal spins, is considered almost correctly. Therefore the exact non-relativistic energy of the system is overestimated by the correlation energy E_{corr}, which is mainly the difference between the exact and the mean repulsion between electrons with opposite spins

$$E_{corr} := E_{exakt} - E_{HF} \leq 0. \qquad (2.43)$$

2.2 QUANTUM MECHANICAL METHODS

In the following, methods as CI, MCSCF, MP2, CC, and DFT will be presented, which account for this correlation energy.

2.2.2 Configuration Interaction

One way to include the correlation energy E_{corr} is the application of a wave function composed of more than one Slater determinant (configuration). The CI trial function $\tilde{\psi}_0$ is represented by a linear combination of configuration state functions (CSFs) Φ_a, which are Slater determinants or linear combinations of a few Slater determinants [42]

$$\tilde{\psi}_0 = \sum_{a=0}^{m} d_a \Phi_a . \tag{2.44}$$

This ansatz accounts for the mixing between the ground state and excited states, and the expansion coefficients d_a can be determined variationally by minimizing the energy expectation value $\tilde{E}_0$ exploiting the variation theorem (2.34) [50]. The constraint of orthonormal CSFs Φ_a yields a usual matrix eigenvalue problem

$$\underline{\underline{H}}\vec{d} = E_{CI}\vec{d} . \tag{2.45}$$

The configurations Φ_a are built by distributing the electrons among the self-consistent spinorbitals ϕ_i obtained by a HF calculation. The number of spinorbitals corresponds to the number of basis functions χ_k, wherefore it is limited. If all possible configurations Φ_a are used for a given finite basis set, the calculation is called full CI [50]. Using an infinite one-particle basis set ($k_{max}=\infty$ in (2.39)) and all corresponding configurations, the full CI method would yield the exact wave function ψ_0 and energy E_0. However, the practical solution for an infinite basis set is not possible, because there are infinite numbers of expansion coefficients c_{ki} and d_a.

Since full CI is only feasible for small molecules and basis sets, one needs criteria, which allow a reasonable choice of CSFs [50]. A systematic approach to the selection of CSFs is to include all those configurations, which differ from the leading or HF configuration Φ_0, which has the largest expansion coefficient d_a, by no more than some spinorbitals. The CSFs are classified by the number of electrons excited from occupied to virtual (unoccupied) spinorbitals of Φ_0, i.e. singles (S) Φ^S, doubles (D) Φ^D, triples (T) Φ^T, ... correspond to configurations, for which one, two, three, ... electrons in occupied spinorbitals were promoted to virtual spinorbitals, respectively. For closed-shell systems Brillouin's theorem indicates that Hamiltonian matrix elements between the leading Φ_0 and singly excited configurations are identically zero ($\langle\Phi_0|\hat{H}|\Phi^S\rangle$=0), i.e. singles do not mix (directly) with the HF configuration. Furthermore, Hamiltonian matrix elements between configurations differing by more than two spinorbitals vanish, i.e. all configurations more than doubly excited do not contribute (directly) to the HF configuration. Thus, a first approach is the limitation of excited configurations to doubles. However, in general singles are also included, because they mix with doubles and thus have a non-zero effect on the calculations. Moreover, some properties as the dipole moment are affected by singles. Therefore in most of the CI calculations the HF configuration as well as singles and doubles are used (CISD).

2.2.3 Multi-configuration Self-consistent Field Method

In contrast to CI in the MCSCF method at least two equivalent CSFs with similar expansion coefficients d_a are used [51]. While in the CI methods the expansion coefficients c_{ki} of (2.39) are determined in a previous HF calculation and held fixed in the CI calculation, both sets of expansion coefficients c_{ki} and d_a of (2.44) are simultaneously optimized in the MCSCF method [50]. The MCSCF method yields a smaller energy eigenvalue than CI, because the expansion coefficients c_{ki} are optimal for the MCSCF and not for the HF wave function. However, the MCSCF method is computationally demanding and often not feasible.

One way to make the MCSCF method efficient is to divide the spinorbitals into active and inactive orbitals [51]. The inactive orbitals composed of the lowest energy spinorbitals are kept doubly occupied in all CSFs, which are built by distributing the remaining active electrons among the active orbitals. If all possible configurations are taken into account, the complete active space SCF (CASSCF) method is obtained.

In this thesis the MCSCF method was used to avoid symmetry-breaking at the orbital level, since the program MOLPRO [52] is limited to the D_{2h} point group and subgroups. If, e.g., a configuration with a single p electron has to be calculated, the electron can either occupy the p_x, p_y, or p_z orbital leading to different spatial symmetries and thus different states ψ_x, ψ_y, and ψ_z [19]. Since the occupation of an orbital lowers its energy, in the case of, e.g., ψ_x only the p_y and p_z orbitals are degenerate, while the p_x orbital is energetically lower. Using the state-averaged MCSCF method the degeneracy of the p orbitals can be described correctly by optimizing the three states ψ_j (j=x, y, z) simultaneously in the same orbital basis. The trial wave functions

2.2 QUANTUM MECHANICAL METHODS

of the three states $\tilde{\psi}_j$ are given as linear combinations of three Slater determinants Φ_i ($i = x, y, z$)

$$\tilde{\psi}_j = \sum_i d_{ij} \Phi_i \ . \tag{2.46}$$

The expansion coefficients d_{ij} of (2.46) and c_{ki} of (2.39) are varied to minimize the energy for the averaged electron density of the three states ψ_j ($j=x, y, z$), whereby each p orbital is occupied by one third.

2.2.4 Coupled Cluster Theory

The CC method account for the mixing between the leading or HF configuration Φ_0 and excited configurations by the following ansatz for the trial wave function [42]

$$\tilde{\psi}_0 = e^{\hat{T}} \Phi_0 := \Phi_0 + \hat{T}\Phi_0 + \frac{1}{2!}\hat{T}^2\Phi_0 + \frac{1}{3!}\hat{T}^3\Phi_0 + ... = \sum_{k=0}^{\infty} \frac{\hat{T}^k}{k!}\Phi_0 \ . \tag{2.47}$$

The exponential operator $e^{\hat{T}}$ is defined by the Taylor-series expansion and the cluster operator $\hat{T}$ is

$$\hat{T} := \hat{T}_1 + \hat{T}_2 + \hat{T}_3 + ... + \hat{T}_n \ , \tag{2.48}$$

where n is the number of electrons in the system, and the one-particle excitation operator $\hat{T}_1$, the two-particle excitation operator $\hat{T}_2$, etc. convert the Slater determinant Φ_0 into linear combinations of all possible singles Φ^S, doubles Φ^D, etc. Since at most n electrons can be excited, no operators beyond $\hat{T}_n$ occur in (2.48). In order to avoid duplication of any excitation, the operators $\hat{T}_i$ generate only determinants with excitations from those spinorbitals, which are occupied in the leading determinant Φ_0 and not from virtual spinorbitals. Thus, $\hat{T}_1^2 \Phi_0$ contains only doubly, and $\hat{T}_2^2 \Phi_0$ only quadruply excited configurations.

Analogous to the full CI method the CC wave function (2.47) is a linear combination of the leading configuration Φ_0 and all possible excitations of the n electrons from occupied to virtual spinorbitals. The aim of both methods CI and CC is to determine the expansion coefficients. However, due to the exponential ansatz the solution of the non-linear CC equations is much more complicated than that of the matrix eigenvalue problem (2.45). Advantages of the CC over the CI method only occur if approximations with respect to the included excitations are taken into account. For example the restriction to doubly excited determinants gives the following CCD and CID wave functions

$$\psi_{CCD} = \Phi_0 + \hat{T}_2\Phi_0 + \frac{1}{2}\hat{T}_2^2\Phi_0 + \frac{1}{6}\hat{T}_2^3\Phi_0 + \ldots \qquad (2.49)$$

and $\quad \psi_{CID} = \Phi_0 + \hat{T}_2\Phi_0\,, \qquad (2.50)$

respectively. While the CI wave function includes only the leading configuration Φ_0 and double excitations Φ^D, the CC wave function contains additionally quadruple, hextuple, etc. excitations beside these configurations. However, the treatment of the higher excitations is only approximate, because the expansion coefficients are optimized for double excitations and the coefficients of, e.g., the quadruply excited determinants are determined as products of these coefficients [42].

From the discussion in Sect. 2.2.2 it is known that the most important excitations are doubles and that singles play an important role for some properties as the dipole moment. Thus, the cluster operator (2.48) is often truncated to include only singles and doubles yielding the so-called CCSD method. Most widely used is the CCSD(T) method, where in addition triples are treated perturbatively.

2.2.5 Second-order Møller–Plesset Perturbation Theory

Perturbation theory is the second major quantum mechanical approximation method besides variation theory. This method is based on the assumption that the Hamiltonian for a true $\hat{H}$ and simpler $\hat{H}^0$ model system differ only slightly by a perturbation $\hat{H}'$ [42]

$$\hat{H} = \hat{H}^0 + \hat{H}'\,. \qquad (2.51)$$

The simpler system with Hamiltonian $\hat{H}^0$ is the so-called unperturbed system

$$\hat{H}^0\psi_s^{(0)} = E_s^{(0)}\psi_s^{(0)}\,, \qquad (2.52)$$

where $E_s^{(0)}$ and $\psi_s^{(0)}$ are the unperturbed energy and wave function of state s, respectively. The task of perturbation theory is to relate the unknown eigenvalues E_s and eigenfunctions ψ_s of the perturbed system

$$\hat{H}\psi_s = E_s\psi_s \qquad (2.53)$$

2.2 QUANTUM MECHANICAL METHODS

to the known eigenvalues and eigenfunctions of the unperturbed system [42]. The Hamiltonian $\hat{H}$ of the true or perturbed system can be written as

$$\hat{H} = \hat{H}^0 + \lambda \hat{H}' . \tag{2.54}$$

Since the Hamiltonian $\hat{H}$ depends on λ, the perturbed wave function ψ_s and energy E_s depend also on λ

$$\psi_s = \psi_s^{(0)} + \lambda \psi_s^{(1)} + \lambda^2 \psi_s^{(2)} + ... \tag{2.55}$$

$$E_s = E_s^{(0)} + \lambda E_s^{(1)} + \lambda^2 E_s^{(2)} + ... \tag{2.56}$$

The unperturbed wave function $\psi_s^{(0)}$ and energy $E_s^{(0)}$ are corrected by terms that are of various orders in the perturbation. The correction terms $\psi_s^{(1)}$ and $E_s^{(1)}$, $\psi_s^{(2)}$ and $E_s^{(2)}$, etc. are called first-order, second-order, etc. correction to the wave function and energy, respectively. For a small perturbation $\hat{H}'$ the consideration of the first few terms of the series will give a good approximation to the true wave function and energy [42]. The application of this method to systems of many interacting particles is called many-body perturbation theory. In 1934 Møller and Plesset proposed a many-body perturbation treatment, where the unperturbed wave function corresponds to the HF function. This form of the many-body perturbation theory is called Møller–Plesset perturbation theory [42].

For the ground state of closed-shell molecules the HF equations (2.35) for electron m in a n-electron molecule are

$$\hat{f}(m)\phi_i(m) = \epsilon_i \phi_i(m) \tag{2.57}$$

$$\text{with} \quad \hat{f}(m) = -\frac{1}{2}\Delta_m - \sum_{I=1}^{N} \frac{Q_I}{r_{mI}} + \sum_{j=1}^{n} \left[\hat{J}_j(m) - \hat{K}_j(m) \right] . \tag{2.58}$$

The Møller–Plesset unperturbed Hamiltonian corresponds to the sum of the one-electron Fock operators $\hat{f}(m)$

$$\hat{H}^0 = \sum_{m=1}^{n} \hat{f}(m) , \tag{2.59}$$

and the HF ground state wave function Φ_0 (2.33) is an eigenfunction of $\hat{H}^0$ with eigenvalue $E_0^{(0)}$ given by the sum of spinorbital energies ϵ_m

$$\hat{H}^0 \Phi_0 = \left(\sum_{m=1}^{n} \epsilon_m \right) \Phi_0 = E_0^{(0)} \Phi_0 . \tag{2.60}$$

The eigenfunctions of the unperturbed Hamiltonian $\hat{H}^0$ are the zero-order wave functions and correspond to all possible Slater determinants formed by distributing the n electrons among the spinorbitals ϕ_i. The perturbation $\hat{H}'$ is the difference between the true $\hat{H}$ and unperturbed $\hat{H}^0$ Hamiltonian (cf. (2.51)) and corresponds to the difference between the true and the HF electron–electron repulsion, which is an average potential [42].

The HF energy E_{HF} is the expectation value associated with the ground state wave function Φ_0 and is equivalent to the sum of the unperturbed energy $E_0^{(0)}$ and its first-order correction $E_0^{(1)}$

$$E_{HF} = \langle\Phi_0|\hat{H}|\Phi_0\rangle = \langle\Phi_0|\hat{H}^0 + \hat{H}'|\Phi_0\rangle = \langle\Phi_0|\hat{H}^0|\Phi_0\rangle + \langle\Phi_0|\hat{H}'|\Phi_0\rangle$$

$$= \langle\psi_0^{(0)}|\hat{H}^0|\psi_0^{(0)}\rangle + \langle\psi_0^{(0)}|\hat{H}'|\psi_0^{(0)}\rangle = E_0^{(0)} + E_0^{(1)}, \quad (2.61)$$

because the unperturbed wave function $\psi_0^{(0)}$ is chosen to be the HF ground state wave function Φ_0. Thus, to improve the HF energy, one has to include at least the second-order energy correction

$$E_0^{(2)} = \sum_{s\neq 0} \frac{\langle\psi_s^{(0)}|\hat{H}'|\Phi_0\rangle\langle\Phi_0|\hat{H}'|\psi_s^{(0)}\rangle}{E_0^{(0)} - E_s^{(0)}} = \sum_{s\neq 0} \frac{\left|\langle\psi_s^{(0)}|\hat{H}'|\Phi_0\rangle\right|^2}{E_0^{(0)} - E_s^{(0)}}. \quad (2.62)$$

Here the unperturbed functions $\psi_s^{(0)}$ are all possible Slater determinants except for Φ_0, i.e. singles Φ^S, doubles Φ^D, ... Therefore the effect of the perturbation is to mix-in contributions from other (excited) states [42].

From the discussion in Sect. 2.2.2 it is known that only the doubly excited determinants Φ^D have non-zero Hamiltonian matrix elements with Φ_0 and thus only doubles contribute to $E_0^{(2)}$. The eigenvalues of doubles Φ^D differ from the eigenvalue of the HF ground state wave function Φ_0 solely by replacement of the energies of two occupied spinorbitals ϵ_i and ϵ_j by those of two virtual spinorbitals ϵ_a and ϵ_b. Hence in (2.62) $E_0^{(0)} - E_s^{(0)} = \epsilon_i + \epsilon_j - \epsilon_a - \epsilon_b$, because the other spinorbitals vanish

$$E_0^{(2)} = \sum_{a<b}^{vir}\sum_{i<j}^{occ} \frac{\left|\langle ab|r_{12}^{-1}|ij\rangle - \langle ab|r_{12}^{-1}|ji\rangle\right|^2}{\epsilon_i + \epsilon_j - \epsilon_a - \epsilon_b}, \quad (2.63)$$

where $\quad \langle ab|r_{12}^{-1}|ij\rangle = \int\int \phi_a^*(1)\phi_b^*(2)r_{12}^{-1}\phi_i(1)\phi_j(2)\mathrm{d}\tau_1\mathrm{d}\tau_2. \quad (2.64)$

2.2 QUANTUM MECHANICAL METHODS

Here ϕ_i and ϕ_j denote occupied and ϕ_a and ϕ_b virtual spinorbitals. Taking the molecular energy as

$$E_0 \approx E_0^{(0)} + E_0^{(1)} + E_0^{(2)} = E_{HF} + E_0^{(2)}, \tag{2.65}$$

gives a MP2 calculation, where the two indicates inclusion of energy corrections up to second-order. In addition to the basis set truncation error of the previous HF calculation, i.e. the usage of an incomplete basis set, an error due to truncation of the Møller–Plesset perturbation energy at $E_0^{(2)}$ occurs.

2.2.6 Density Functional Theory

The foundation for DFT was created by Hohenberg and Kohn in 1964 by proving that the electron density $\rho_0(x, y, z)$ of the ground state contains all necessary informations to determine the ground state energy, wave function, and all other electronic properties (Hohenberg–Kohn theorem) [49]. Thus, the ground state electronic energy E_0 is a functional of ρ_0

$$E_0 = E_0[\rho_0], \tag{2.66}$$

where the square brackets denote a functional relation [42]. The electron probability density ρ is the probability of finding any of the n electrons (with arbitrary spin σ) within the volume $d\vec{r}_1$ and is defined as

$$\rho(x,y,z) = \rho(\vec{r}_1) = n \int ... \int \psi^*\psi \mathrm{d}\sigma_1 \mathrm{d}\tau_2 ... \mathrm{d}\tau_n \tag{2.67}$$

with $\quad \mathrm{d}\tau_i := \mathrm{d}\vec{r}_i \mathrm{d}\sigma_i \quad$ and $\quad \psi = \psi(\vec{r}_1, \vec{r}_2, ..., \vec{r}_n, \sigma_1, \sigma_2, ..., \sigma_n)$. (2.68)

The purely electronic Hamiltonian $\hat{H}$ is given by

$$\hat{H} = \hat{T} + \hat{V}_{eN} + \hat{V}_{ee} = -\frac{1}{2}\sum_{i=1}^{n}\Delta_i - \sum_{i=1}^{n}\sum_{I=1}^{N}\frac{Q_I}{r_{iI}} + \sum_{i<j}^{n}\frac{1}{r_{ij}}. \tag{2.69}$$

The kinetic energy of the electrons $\hat{T}$ as well as the electron–electron repulsion $\hat{V}_{ee}$ are represented by universal operators depending only on the number of electrons n in the system. The electron–nucleus attraction $\hat{V}_{eN}$ is a specific operator depending on the atoms in the system, i.e. on the number of nuclei N, their positions $\vec{R}_I$, and their charges Q_I. It is also called external potential, since it is produced by charges external to the system of electrons [42].

The electron probability density ρ includes all informations for the electronic Hamiltonian $\hat{H}$. The number of electrons n can be calculated by

$$\int \rho(\vec{r}_1)\mathrm{d}\vec{r}_1 = n \qquad (2.70)$$

using the normalization of ψ. The informations on the nuclei can be determined from the maxima of the electron probability density ρ, i.e. the number of nuclei N is equal to the number of maxima, the nuclear positions $\vec{R}_I$ correspond to the maxima positions, and the nuclear charges Q_I are proportional to the gradient at the maxima positions. Since the ground state energy E_0 is a unique functional of ρ_0, so must be its individual parts [42]

$$E_0[\rho_0] = \hat{T}[\rho_0] + \hat{V}_{eN}[\rho_0] + \hat{V}_{ee}[\rho_0] \ . \qquad (2.71)$$

While the functionals $\hat{T}[\rho_0]$ and $\hat{V}_{ee}[\rho_0]$ are unknown, the electron–nucleus attraction $\hat{V}_{eN}[\rho_0]$ is known and given as [42]

$$\hat{V}_{eN} = -\left\langle \psi_0 \left| \sum_{i=1}^{n} \sum_{I=1}^{N} \frac{Q_I}{r_{iI}} \right| \psi_0 \right\rangle = -\sum_{I=1}^{N} Q_I \int \frac{\rho_0(\vec{r}_1)}{r_{1I}} \mathrm{d}\vec{r}_1 \ . \qquad (2.72)$$

In 1965 Kohn and Sham devised a practical method to find the electron probability density ρ_0 and to calculate the ground state energy E_0 from this electron density, i.e. they provided a solution to the problem of the unknown energy terms $\hat{T}[\rho_0]$ and $\hat{V}_{ee}[\rho_0]$ [53]. They suggested to formally split the functionals $\hat{T}[\rho_0]$ and $\hat{V}_{ee}[\rho_0]$ into two parts, where one part $\hat{T}^{KS}[\rho_0]$ and $\hat{V}_{ee}^{KS}[\rho_0]$ can be exactly calculated and the other still unknown part $\Delta\hat{T}[\rho_0]$ and $\Delta\hat{V}_{ee}[\rho_0]$ contains the deviation of these terms from the real functionals [42]

$$\Delta\hat{T}[\rho_0] = \hat{T}[\rho_0] - \hat{T}^{KS}[\rho_0] \qquad (2.73)$$

$$\Delta\hat{V}_{ee}[\rho_0] = \hat{V}_{ee}[\rho_0] - \hat{V}_{ee}^{KS}[\rho_0] \ . \qquad (2.74)$$

The electron–electron interaction $\hat{V}_{ee}^{KS}[\rho_0]$, which can be determined exactly, is the classical expression for the electrostatic electron–electron repulsion energy, if the electrons were smeared out into a continuous distribution of charge with electron density ρ_0 [42]

$$\hat{V}_{ee}^{KS}[\rho_0] = -\frac{1}{2} \int \int \frac{\rho_0(\vec{r}_1)\rho_0(\vec{r}_2)}{r_{12}} \mathrm{d}\vec{r}_1 \mathrm{d}\vec{r}_2 \ . \qquad (2.75)$$

2.2 QUANTUM MECHANICAL METHODS

For a reference system of n non-interacting electrons the Hamiltonian $\hat{H}^{KS}$ is

$$\hat{H}^{KS} = \sum_{i=1}^{n}\left[-\frac{1}{2}\Delta_i + \sum_{I=1}^{N}\frac{Q_I}{r_{iI}}\right] = \sum_{i=1}^{n}\hat{h}_i^{KS} \qquad (2.76)$$

and the kinetic energy $\hat{T}^{KS}[\rho_0]$ can be calculated by

$$\hat{T}^{KS}[\rho_0] = -\frac{1}{2}\sum_{i=1}^{n}\langle\phi_i^{KS}(1)|\Delta_1|\phi_i^{KS}(1)\rangle, \qquad (2.77)$$

where each Kohn–Sham spinorbital ϕ_i^{KS} is an eigenfunction of the one-electron operator $\hat{h}_i^{KS}$ [42].
As mentioned before the functionals $\Delta\hat{T}[\rho_0]$ and $\Delta\hat{V}_{ee}[\rho_0]$ are unknown and contain the deviation from the real system. Defining the exchange–correlation energy functional $E_{xc}[\rho_0]$ by [42]

$$E_{xc}[\rho_0] := \Delta\hat{T}[\rho_0] + \Delta\hat{V}_{ee}[\rho_0], \qquad (2.78)$$

the energy of the ground state can be written as

$$E_0[\rho_0] = -\frac{1}{2}\sum_{i=1}^{n}\langle\phi_i^{KS}(1)|\Delta_1|\phi_i^{KS}(1)\rangle - \sum_{I=1}^{N}Q_I\int\frac{\rho_0(\vec{r}_1)}{r_{1I}}d\vec{r}_1$$

$$+\frac{1}{2}\int\int\frac{\rho_0(\vec{r}_1)\rho_0(\vec{r}_2)}{r_{12}}d\vec{r}_1 d\vec{r}_2 + E_{xc}[\rho_0]. \qquad (2.79)$$

The first three terms on the right side of this equation are easy to evaluate from ρ_0 and include the main contributions to the ground state energy. The fourth quantity E_{xc}, although not easy to evaluate exactly, will be a relatively small term.

Hence, the ground state energy E_0 can be calculated from ρ_0, if the Kohn–Sham spinorbitals ϕ_i^{KS} are found and if the functional E_{xc} is known. The spinorbitals ϕ_i^{KS} can be determined by using the variational theorem (2.34) [42]. However, the exchange–correlation functional E_{xc} has to be approximated. Numerous schemes were developed to obtain approximate forms for the exchange–correlation functional, e.g. the local density approximation based on a uniform electron gas, the generalized gradient approximation, which takes into account that the electron distribution within a molecule is far from being uniform, or hybrid-type functionals, which mix together the exact HF exchange E_x with gradient-corrected E_{xc} formulas [42].

Chapter 3

Results and Discussion

3.1 5f-in-core Pseudopotentials for Actinides

In this Section the adjustment of the quasirelativistic energy-consistent 5f-in-core PPs for di- [29], tetra- [29], penta- [30], and hexavalent [30] actinides will be described. Furthermore, the adjustment of CPPs for the di- [29], tri- [54], and tetravalent [29] PPs and the optimization of valence basis sets [29, 30] will be presented. Finally, atomic and molecular test calculations as well as some applications will be discussed, in order to assess the accuracy of the PPs, CPPs, and basis sets, respectively.

3.1.1 Adjustment of the Pseudopotentials

The 5f-in-core PPs corresponding to di- ($5f^{n+1}$, $n=5$–13 for Pu–No)[1] [29], tetra- ($5f^{n-1}$, $n=1$–9 for Th–Cf) [29], penta- ($5f^{n-2}$, $n=2$–6 for Pa–Am) [30], and hexavalent ($5f^{n-3}$, $n=3$–6 for U–Am) [30] actinide atoms were similarly generated as the PPs corresponding to trivalent oxidation states ($5f^n$, $n=0$–14 for Ac–Lr) [20]. The parameters for di- and tetra- as well as for penta- and hexavalent PPs are listed in Tables A.3 and A.4, respectively. The 1s–5f (spherically averaged) shells are included in the PP core, while all orbitals with main quantum number larger than five are treated explicitly, i.e. 10, 12, 13, and 14 valence electrons for the di-, tetra-, penta-, and hexavalent PPs, respectively. In the case of the di- and tetravalent PPs the s-, p-, and d-PPs are composed of two Gaussians each and were adjusted by a least-squares fit to the total valence energies of nine and 18 reference states, respectively (cf. Table A.1). In the case of the

---
[1] The parameters of the divalent 5f-in-core PPs I have already adjusted during my diploma thesis [19].

penta- and hexavalent PPs the s-PPs are composed of three and the p- and d-PPs of two Gaussians, which were adjusted to 18 reference states (cf. Table A.2). The reference data were taken from relativistic AE calculations using the WB scalar-relativistic HF approach. Both AE WB as well as PP calculations were performed with an atomic finite-difference HF scheme [55].

In order to allow for some participation of the 5f orbitals in chemical bonding, the f-parts of the PPs are designed to describe partial occupations of the 5f shell, which are larger than the integral occupation number implied by the valency, i.e. $5f^{n+1+q}$ (n=5–13 for Pu–No), $5f^{n-1+q}$ (n=1–9 for Th–Cf), $5f^{n-2+q}$ (n=2–6 for Pa–Am), and $5f^{n-3+q}$ (n=3–6 for U–Am) with $0 \leq q < 1$ for di-, tetra-, penta-, and hexavalent actinide atoms, respectively [18]. These f-PPs consist of two types of potentials V_1 and V_2 which are linearly combined as follows [20]

$$V = \left(1 - \frac{m}{14}\right) V_1 + \frac{m}{14} V_2 . \qquad (3.1)$$

Here m is the integral number of electrons in the 5f orbitals kept in the core, i.e. m=n+1, m=n−1, m=n−2, and m=n−3 for the di-, tetra-, penta-, and hexavalent case, respectively. V_1 and V_2 model 5f shells, which can and cannot accommodate an additional electron, respectively. Thus, V_1 is the exact potential for a $5f^0$ occupation, whereas V_2 is exact for $5f^{14}$. V_1 was adjusted to four reference configurations of highly-charged actinide ions (An^{n+}, n=9, 11–13 for di-, tetra-, penta-, and hexavalent PPs), where a single valence electron is situated in the 5f, 6f, 7f, and 8f shell, respectively. For V_2 only three reference configurations were used, i.e. the reference configuration with the additional electron in the 5f shell was omitted.

For divalent PPs the errors in the total valence energies of finite-difference HF calculations are smaller than 0.07 and 0.10 eV for s-, p-, d-parts and f-parts, respectively (cf. Tables A.5 and A.7). For tetravalent PPs these errors are smaller than 0.10 and 0.15 eV (cf. Tables A.5 and A.7). For pentavalent PPs these errors are below 0.06 and 0.30 eV, and for hexavalent PPs below 0.03 and 0.76 eV, respectively (cf. Tables A.6 and A.8). Since the errors for the penta- and hexavalent f-PPs are clearly larger than 0.1 eV, it was tried to use two Gaussians for V_1 and V_2. In this way the deviations could be reduced to at most 0.05 eV for both penta- and hexavalent PPs. However, these f-PPs yield too strong 5f orbital participations especially in the case of $5f^0$, i.e. for PaF$_5$ and UF$_6$ LPP 5f occupations are by 0.15 and 0.52 electrons larger than SPP 5f occupations, respectively (5f occupations for LPP/SPP: PaF$_5$ 0.71/0.56; UF$_6$ 1.68/1.16 electrons). Moreover, the deviations between LPP and SPP bond lengths and energies are at least twice as large, if two Gaussians instead of one are used for V_1 and V_2 (f-PP with one/two Gaussians: PaF$_5$ ΔR_{ax}=0.013/0.032 Å, ΔR_{eq}=0.012/0.039 Å, ΔE=0.052/0.334 eV; UF$_6$ ΔR=0.0004/0.042 Å, ΔE=0.29/0.68 eV). Thus, only one Gaussian is used for V_1 and V_2 and the greater errors of 0.30 (0.4%) and 0.76 eV (0.7%) for the total valence energies of the highly-charged ions An^{12+} and An^{13+} for the adjustment of penta- and hexavalent f-PPs are accepted, to avoid a too strong 5f orbital participation.

3.1.2 Adjustment of the Core-polarization Potentials

The parameters of the CPPs, i.e. the dipole polarizabilities and cutoff parameters, are listed in Table A.10. For the CPPs corresponding to di- [29], tri- [54], and tetravalent [29] 5f-in-core PPs the DHF dipole polarizabilities of Ra^{10+} (1.0407 a.u.) and No^{10+} (6.4819 a.u.), Ac^{11+} (0.8982 a.u.) and Lr^{11+} (3.7501 a.u.), and Th^{12+} (0.7830 a.u.) and Rf^{12+} (2.5179 a.u.) were used to interpolate those of the other An^{10+}, An^{11+}, and An^{12+} cores, respectively, because the DHF program package [56] can only handle closed-shell systems. Since the dipole polarizabilities of the highly-charged PP cores An^{10+}, An^{11+}, and An^{12+} are strongly dependent on the presence of the valence electrons, the polarizabilities were calculated using the orbitals of the neutral An atoms, the An$^+$, and the An^{2+} cations with the subconfiguration $6s^26p^67s^2$ for di-, tri-, and tetravalent CPPs, respectively.

Since the MOLPRO program package [52] provides two possible cutoff factors

$$\omega(r) = \left(1 - \exp\left(-\delta r^2\right)\right)^{0.5} \quad (ntype=1 \text{ in MOLPRO}) \tag{3.2}$$

and

$$\omega(r) = 1 - \exp\left(-\delta r^2\right) \quad (ntype=2 \text{ in MOLPRO}), \tag{3.3}$$

first it was tested, which cutoff factor yields the best results in the case of the 5f-in-core LPPs. Some electronic transitions, where a 7s electron is ionized, were calculated at the LPP+CPP CCSD(T) level using the two different cutoff factors, and the results were compared to SPP CCSD(T) calculations, because no or not enough experimental data are available. The SPP CCSD(T) data were calculated without SO coupling at the basis set limit [21] except for the transitions using the tetravalent PPs, which were calculated using the standard basis sets (14s13p10d8f6g)/[6s6p5d4f3g] [21]. The LPP+CPP CCSD(T) calculations were carried out with MOLPRO [52] using

(10s10p10d8f6g) even-tempered basis sets, which were CCSD(T) energy-optimized for the $6d^17s^1$, $6d^17s^2$ (Ac)/$7s^27p^1$ (Lr), and $6d^27s^2$ valence subconfigurations of the neutral atoms for di-, tri-, and tetravalent PPs, respectively. In both LPP+CPP and SPP CCSD(T) calculations no orbitals were frozen. In the case of the divalent PPs the cutoff parameter for both cutoff factors was chosen to be $\delta=0.5$, while for the tri- and tetravalent PPs it was chosen to be $\delta=1.0$. As one can see from Table 3.1 the SPP values are in general overestimated by up to 1.45 (Lr) and 0.35 eV (first transition of Bk) for *ntype=1* and *ntype=2*, respectively. Only for the second transitions of Am and Th the LPP+CPP calculation using *ntype=2* and both LPP+CPP calculations underestimate the SPP values, respectively. Since *ntype=2* yields always smaller values than *ntype=1* except for the second transition of Th, the results using *ntype=2* agree clearly better with the SPP reference data. Therefore this cutoff factor was chosen for the CPPs corresponding to the actinide 5f-in-core LPPs. In contrast to this for the lanthanide 4f-in-core LPPs the CPP cutoff factor with *ntype=1* is used [41]. This can be explained by the fact that for lanthanides experimental values are available, which are typically underestimated by quantum mechanical calculations. Thus, in comparison to experimental data the larger values obtained by using *ntype=1* should be favored.

Table 3.1: Electronic transitions (in eV) calculated at the LPP+CPP CCSD(T) level using both possible cutoff factors *ntype=1* (3.2) and *ntype=2* (3.3) in comparison to SPP CCSD(T) reference data [21].

An	Transition	SPP	*ntype=1*	*ntype=2*
divalent CPPs				
Am	$7s^2 \rightarrow 7s^1$	5.85	6.07	5.95
	$7s^1 \rightarrow 7s^0$	11.81	11.87	11.61
No	$7s^2 \rightarrow 7s^1$	6.50	7.02	6.55
	$7s^1 \rightarrow 7s^0$	12.69	13.79	12.84
trivalent CPPs				
Ac	$7s^2 \rightarrow 7s^1$	11.73	11.78	11.75
Lr	$7s^2 \rightarrow 7s^1$	14.40	15.85	14.68
tetravalent CPPs				
Th	$6d^27s^2 \rightarrow 6d^27s^1$	6.29	6.30	6.29
	$6d^27s^1 \rightarrow 6d^27s^0$	12.35	12.13	12.30
Bk	$6d^27s^2 \rightarrow 6d^27s^1$	6.68	7.16	7.03
	$6d^27s^1 \rightarrow 6d^27s^0$	13.78	14.13	13.88

The cutoff parameters were fitted at the CCSD(T) level to the following IPs: IP_1+IP_2 of Am and IP_1, IP_2 of No (divalent PPs); IP_1, IP_2 of Ac and IP_2, IP_3 of Lr (trivalent

3.1   5F-IN-CORE PSEUDOPOTENTIALS FOR ACTINIDES   39

PPs); IP_1, IP_2 of Th and IP_2, IP_3 of Bk (tetravalent PPs). The reason why these actinide elements were chosen are their unoccupied (Ac, Th), half occupied (Am, Bk), and fully occupied (No, Lr) 5f orbitals, respectively, since in these cases more accurate reference data are available, i.e. SPP CCSD(T) calculations without SO coupling using extrapolation to the basis set limit [21]. However, for the IPs of Th and Bk needed to adjust the tetravalent CPPs standard basis sets [21] were taken to obtain the SPP CCSD(T) reference data. The LPP+CPP CCSD(T) calculations were carried out with MOLPRO [52] using the (10s10p10d8f6g) even-tempered basis sets. The cutoff parameters of the other di-, tri-, and tetravalent actinide elements were interpolated using the values of Am (0.6980)/No (0.2404), Ac (0.8727)/Lr (0.2696), and Th (0.9293)/Bk (0.4867), respectively.

3.1.3   Optimization of the Valence Basis Sets

In the following the optimization of the valence basis sets corresponding to di- [29], tetra- [29], penta- [30], and hexavalent [30] 5f-in-core PPs will be presented in two parts, i.e. first the generation of the primitive and then that of the segmented contracted basis sets will be given. The basis set parameters are listed in Tables B.1–B.4 for di-, tetra-, penta-, and hexavalent PPs, respectively.

3.1.3.1   Primitive Basis Sets

The Gaussian-type-orbital (GTO) valence basis sets for di- [29], tetra- [29], penta- [30], and hexavalent [30] 5f-in-core PPs were constructed analogous to those for trivalent PPs [20]. But here only two different sets of primitive Gaussian functions (4s4p3d)+2s1p1d and (5s5p4d)+2s1p1d were derived, since the (6s6p5d)+2s1p1d basis sets for trivalent PPs yield results, which are almost of the same quality as those of the (5s5p4d)+2s1p1d basis sets.

First, basis sets for use in crystal calculations were created, i.e. in the divalent case (4s4p) and (5s5p) basis sets were HF energy-optimized [57] for the An^{2+} $6s^26p^6$ valence subconfiguration. In the tetra-, penta-, and hexavalent case (4s4p3d) and (5s5p4d) basis sets were HF energy-optimized [57] for the $6s^26p^66d^1$ valence subconfiguration of An^{3+}, An^{4+}, and An^{5+}, respectively. To avoid linear dependency in solid state calculations, which are usually caused by overlap between too diffuse functions of the densely packed atoms, all exponents, which became smaller than 0.15, were fixed to this value and the remaining exponents were reoptimized. Furthermore, all optimizations were carried out with the requirement that the ratio of exponents in the same angular symmetry must be at least 1.5, because in particular steep Gaussians tend to a coalescence resulting in a linearly dependent basis [58]. For divalent PPs the basis set errors in the valence energies with respect to numerical finite-difference LPP HF calculations [55] are below 0.11 and 0.03 eV for (4s4p) and (5s5p), respectively, and for tetravalent PPs these errors are below 0.15 and 0.07 eV for (4s4p3d) and (5s5p4d), respectively (cf. Table B.5). For pentavalent PPs these errors are below 0.09 and 0.02 eV for (4s4p3d) and (5s5p4d), respectively, and for hexavalent PPs these errors are smaller than 0.13 and 0.03 eV (cf. Table B.6). Thus, at least the larger (5s5p) and (5s5p4d) basis sets show errors below 0.1 eV, which is the requested accuracy, because it corresponds to the accuracy of the PP adjustment.

Secondly, the valence basis sets were augmented by adding a set of 2s1p4d and 2s1p5d low-exponent Gaussians to (4s4p) and (5s5p), respectively, as well as a set of 2s1p1d to (4s4p3d) and (5s5p4d) yielding final (6s5p4d) and (7s6p5d) primitive sets for use in molecular calculations. The added exponents were HF energy-optimized [57] for the $7s^2$ (^1S, s-basis), $7s^1 7p^1$ (^3P, p-basis), and $6d^1 7s^1$ (^3D, d-basis) valence substates of the neutral actinides for the divalent PPs as well as for the $6d^2 7s^2$ valence subconfiguration of the neutral actinides for the tetravalent PPs. In the case of the penta- and hexavalent PPs the additional exponents were HF energy-optimized [57] for the $6d^3 7s^2$ and the $6d^4 7s^2$ valence subconfigurations of the neutral actinides, respectively. The differences in the valence energies with respect to numerical finite-difference LPP HF calculations [55] are below 0.15 and 0.04 eV for (6s5p4d) and (7s6p5d) of divalent PPs, respectively, and below 0.14 and 0.08 eV for tetravalent PPs (cf. Table B.5). For pentavalent PPs these differences are below 0.11 and 0.03 eV for (6s5p4d) and (7s6p5d), respectively, and for hexavalent PPs below 0.16 and 0.05 eV (cf. Table B.6). Therefore also the (7s6p5d) molecular basis sets are clearly below 0.1 eV.

Finally, sets of 2f1g correlation/polarization functions were energy-optimized in CI calculations [52] for the $7s^2$, $6d^2 7s^2$, $6d^3 7s^2$, and $6d^4 7s^2$ ground state valence subconfigurations for di-, tetra-, penta-, and hexavalent PPs, respectively. For penta- and hexavalent PPs all exponents were optimized explicitly. However, for di- and tetravalent PPs only the exponents of Pu, Fm–No and Th, Pa, Bk were optimized explicitly, while those of Am–Es and of U–Cm, Cf were interpolated.

3.1.3.2 Contracted Basis Sets

The basis sets were contracted using different segmented contraction schemes (cf. Table 3.2) to yield basis sets of approximately valence double-, triple-, and quadruple-zeta (VDZ, VTZ, and VQZ) quality for the s and p symmetries. In case of d symmetry at least a triple-zeta contraction was necessary and additional sets with less tight d contraction are also offered (VDZ: [4s3p3d], VTZ: [5s4p3d], [5s4p4d], and VQZ: [6s5p4d]).

Table 3.2: Contraction schemes.

Contraction	Contraction Scheme	
	(6s5p4d)	(7s6p5d)
[4s3p3d]	{2211/221/211}	{3211/321/311}
[5s4p4d]^a	{21111/2111/211}	{31111/3111/2111}
[6s5p4d]		{211111/21111/2111}

^{a}In the case of the (6s5p4d) basis set the VTZ contraction is [5s4p3d].

The errors in total valence energies of the $6d^17s^1$ (divalent) and $6d^27s^2$ (tetravalent) valence substates with respect to numerical finite-difference LPP HF calculations [55] of all contracted basis sets are below 0.2 eV (cf. Tables B.7 and B.8). In the case of the divalent PPs all contractions of the (7s6p5d) and for the tetravalent case the VQZ contraction of (7s6p5d) yield errors smaller than 0.1 eV. For pentavalent PPs the errors in total valence energies of the $6d^37s^2$ valence substates for the contracted (6s5p4d) basis sets, the VDZ as well as the VTZ contracted (7s6p5d) basis sets, and the VQZ contracted (7s6p5d) basis sets are below 0.17, 0.08, and 0.03 eV, respectively (cf. Table B.9). For hexavalent PPs these errors in total valence energies of the $6d^47s^2$ valence substates are smaller than 0.33, 0.13, and 0.06 eV, respectively (cf. Table B.10). Thus, at least the (7s6p5d)/[6s5p4d] VQZ basis sets fulfill the requested accuracy of 0.1 eV.

3.1.4 Atomic Test Calculations

In order to test the di- and tetravalent 5f-in-core LPPs [29] and their corresponding basis sets as well as the di- [29], tri- [54], and tetravalent [29] CPPs, the first and second IPs for the actinides were calculated. Table 3.3 lists the electronic ground states and the configurations of the singly- and doubly-charged actinides. The 5f occupation of the configurations determine which LPP has to be used to calculate the IPs, i.e. $5f^{n+1}$,

$5f^n$, or $5f^{n-1}$ (n=0–14 for Ac–Lr) occupations correspond to di-, tri-, and tetravalent oxidation states, respectively. As one can see from Table 3.3 for eight, six, and one actinide di-, tri-, and tetravalent LPPs were used, respectively. In the case of the second IPs the values of Th, U, Np, and Cm could not be calculated at the LPP level, because for these IPs the 5f occupation changes due to the ionization, what cannot be described, if the 5f shell is included in the PP core.

Table 3.3: Electronic ground states and configurations for the actinide cations An^{n+} (n=1–2) [7]. Configurations with 5f occupations corresponding to di-, tri-, and tetravalent oxidation states are given in black, in italics, and underlined, respectively. Configurations, whose 5f occupation changes due to the ionization, are given in bold.

An	An	An^{1+}	An^{2+}
Ac	$\underline{6d^1 7s^2}$	$7s^2$	$7s^1$
Th	$\underline{6d^2 7s^2}$	$\underline{6d^2 7s^1}$	**$5f^1 6d^1$**
Pa	*$5f^2 6d^1 7s^2$*	*$5f^2 7s^2$*	*$5f^2 6d^1$*
U	*$5f^3 6d^1 7s^2$*	*$5f^3 7s^2$*	**$5f^4$**
Np	*$5f^4 6d^1 7s^2$*	*$5f^4 6d^1 7s^1$*	**$5f^5$**
Pu	$5f^6 7s^2$	$5f^6 7s^1$	$5f^6$
Am	$5f^7 7s^2$	$5f^7 7s^1$	$5f^7$
Cm	*$5f^7 6d^1 7s^2$*	*$5f^7 7s^2$*	**$5f^8$**
Bk	$5f^9 7s^2$	$5f^9 7s^1$	$5f^9$
Cf	$5f^{10} 7s^2$	$5f^{10} 7s^1$	$5f^{10}$
Es	$5f^{11} 7s^2$	$5f^{11} 7s^1$	$5f^{11}$
Fm	$5f^{12} 7s^2$	$5f^{12} 7s^1$	$5f^{12}$
Md	$5f^{13} 7s^2$	$5f^{13} 7s^1$	$5f^{13}$
No	$5f^{14} 7s^2$	$5f^{14} 7s^1$	$5f^{14}$
Lr	*$5f^{14} 7s^2 7p^1$*	*$5f^{14} 7s^2$*	*$5f^{14} 7s^1$*

State-averaged MCSCF with subsequent CCSD(T) calculations were performed in MOLPRO [52] using LPPs with and without CPPs and uncontracted (7s6p5d2f1g) valence basis sets. The state-averaging was necessary to avoid symmetry-breaking at the orbital level, since MOLPRO [52] is limited to the D_{2h} point group and subgroups. In the CCSD(T) calculations no orbitals were frozen. Since experimental data [59–61] are only available for IP_1 (except for Lr) and IP_2 of Ac, the LPP results are also compared to 5f-in-valence SPP calculations from the literature [21]. The SPP IPs were determined using state-averaged CASSCF with subsequent multi-reference averaged coupled-pair functional (ACPF) calculations [52]. In the ACPF calculations the 5s, 5p,

3.1 5F-IN-CORE PSEUDOPOTENTIALS FOR ACTINIDES

and 5d orbitals were frozen. The calculations did not include SO coupling and were performed at the basis set limit, i.e. (14s13p10d8f6g6h6i) basis sets were used except for Pa, where the standard basis set (14s13p10d8f6g)/[6s6p5d4f3g] had to be applied due to convergence problems. The LPP results as well as the experimental [59–61] and SPP reference data are listed in Table C.1.

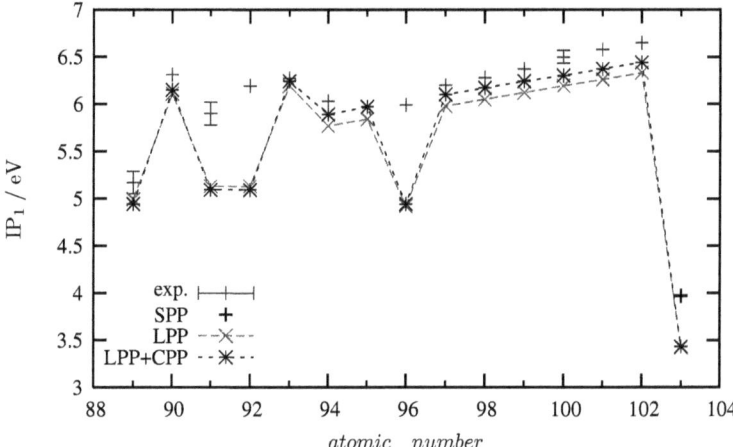

Figure 3.1: First ionization potentials IP_1 (in eV) for the actinides from LPP CCSD(T) calculations [52] with and without CPPs in comparison to experimental data [59–61] except for Lr, where a SPP multi-reference ACPF value without SO coupling at the basis set limit [21] is given. In the LPP calculations (7s6p5d2f1g) basis sets were applied.

Figure 3.1 shows the first ionization potentials for the actinides from LPP calculations with and without CPPs in comparison to experimental data except for Lr, where a SPP value is given. The experimental data are always underestimated by both LPP and LPP+CPP calculations. The mean absolute error (m.a.e.) and mean relative error (m.r.e.) for the LPP data amount to 0.40 eV and 6.8%, respectively (cf. Table C.1). The deviations are smaller than 0.35 eV except for Pa, U, Cm, and Lr, where they are 0.55–1.07 eV corresponding to 14–18%. One reason for these large differences are the neglect of SO effects. However, the SO effects calculated at the SPP level [21] amount only to 0.10, 0.14, and 0.18 eV for Pa, U, and Cm, respectively, and the reference value for Lr is a SPP ACPF calculation without SO coupling. Therefore the SO coupling explains only a small part of these errors. The more crucial reason is that the IPs calculated at the 5f-in-core LPP level occur between averaged instead of high-spin LS-states, because the 5f shell is included in the PP core and is thus treated in an averaged manner. In Fig. 3.2 the SPP/experimental and LPP transitions for $5f^n6d^17s^2$ to $5f^n7s^2$ are given. As one can see the gap between the high-spin and low-spin states is the larger the more unpaired electrons exist in the considered configuration, i.e. the splitting for the initial $5f^n6d^17s^2$ is clearly bigger than that for the final $5f^n7s^2$ state. Therefore the LPP IP is too small, because the energy loss for the initial state is bigger than the energy gain for the final state. Since for Pa, U, Cm, and Lr a 6d or 7p electron is ionized, the experimental or SPP values are underestimated. For the other elements this error does not occur, because a 7s electron is ionized. For the first IP of uranium, i.e. $5f^36d^17s^2$ to $5f^37s^2$, AE WB calculations [55] show that the ionization between averaged states (IP_1=4.53 eV) are by 0.92 eV smaller than that between high-spin states (IP_1=5.45 eV), i.e. ^5L to ^4I. If one adds this amount to the LPP value for the IP_1 of U (IP_1=5.13 eV), the remaining deviation from the experimental value (IP_1=6.19 eV) is 0.14 eV and can be explained by the missing SO coupling.

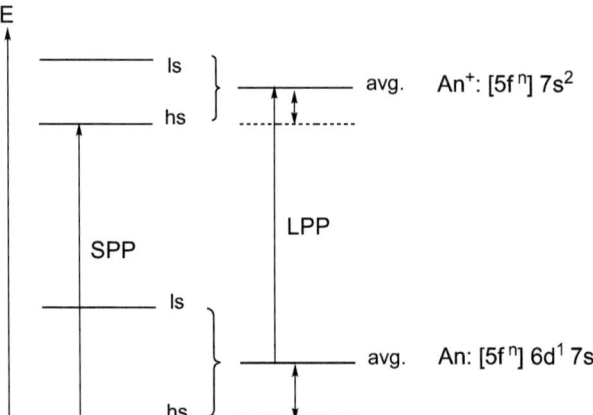

Figure 3.2: Experimental or SPP ionizations occur between high-spin (hs) states, while LPP ionizations take place between averaged (avg.) LS-states, because the 5f shell is included in the PP core and is thus treated in an averaged manner. The gap between the high-spin and low-spin (ls) states is the larger the more unpaired electrons exist in the considered configuration.

3.1 5F-IN-CORE PSEUDOPOTENTIALS FOR ACTINIDES

If CPPs are applied, the IPs are increased except for Ac, Pa, and U, where they are by at most 0.05 eV smaller than the pure LPP results. The CPP effect decreases from di- via tri- to tetravalent oxidation states, i.e. the changes of the IPs amount to 0.11–0.13, 0.01–0.05, and 0.04 eV for di-, tri-, and tetravalent CPPs, respectively. The reason for this decrease is that the CPP is proportional to the dipole polarizability (cf. (2.1)), which is the higher the larger the ionic radius. Since the ionic radius is in turn the larger the smaller the ionic charge, the dipole polarizability becomes smaller with increasing ionic charge, e.g., for Cm the dipole polarizability decreases from Cm^{10+} 4.1500 via Cm^{11+} 2.3242 to Cm^{12+} 1.5265 a.u. (cf. Table A.10).

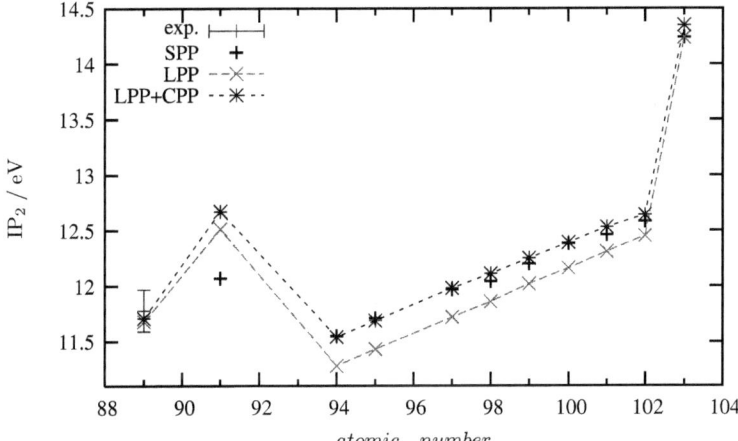

Figure 3.3: Second ionization potentials IP_2 (in eV) for the actinides from LPP CCSD(T) calculations [52] with and without CPPs in comparison to SPP multi-reference ACPF data without SO coupling at the basis set limit [21] except for Ac, where an experimental value [59–61] is given. In the LPP calculations (7s6p5d2f1g) basis sets were applied, and for Pa only the standard basis set (14s13p10d8f6g)/[6s6p5d4f3g] was used in the SPP calculation.

Because the first IPs are underestimated using pure LPPs, the application of CPPs improves the results except for Ac, Pa, and U, i.e. the m.a.e. and m.r.e. are 0.33 eV and 5.8%, respectively. The deviations are smaller than 0.25 eV except for Pa, U, Cm, and Lr, where they amount to at most 1.10 eV (18%). Thus, in summary the CPPs improve the LPP results especially for the divalent oxidation state, where they show the largest effect due to the high dipole polarizability.

Figure 3.3 shows the second ionization potentials for the actinides from LPP calculations with and without CPPs in comparison to SPP data except for Ac, where an experimental value is available. The reference data are underestimated by the LPP results except for Pa, where it is overestimated by 0.44 eV. The m.a.e. and m.r.e. amount to 0.20 eV and 1.7%, respectively, and are half as large as those of the first IPs. This is due to the fact that the SPP data are calculated without SO coupling and that only for Pa the IP includes a change of the 6d occupation, i.e. the transition goes from $5f^2 7s^2$ to $5f^2 6d^1$. Since here the final state consists of more unpaired electrons, the IP is overestimated.

The use of CPPs increases the IPs in a range of 0.19–0.26 and 0.03–0.16 eV for di- and trivalent oxidation states, respectively. Due to this increase the m.a.e. (m.r.e.) is reduced by 50% to 0.10 eV (0.8%). Therefore the CPPs are (at least in atomic calculations) useful tools to improve the accuracy of the LPP results.

3.1.5 Molecular Test Calculations

For the 5f-in-core LPPs as well as corresponding CPPs and valence basis sets molecular test calculations were performed for actinide fluorides AnF_n (n=2–6). In this section first the computational details of these test calculations will be given. Next the results of the di- and tetravalent LPPs, CPPs, and basis sets [29] will be discussed. Then those of the penta- and hexavalent LPPs and basis sets [30] will be presented. Subsequently, the molecular tests of the trivalent CPPs [54], will be given. Finally, the results for all AnF_n (n=2–6) will be compared to each other.

3.1.5.1 Computational Details

The test calculations[2] for AnF_2 (An=Pu–No) and AnF_4 (An=Th–Cf) were carried out with MOLPRO [52] using the di- and tetravalent 5f-in-core LPPs [29] with and without CPPs. For AnF_5 (An=Pa–Am) and AnF_6 (An=U–Am) analogous test calculations using the penta- and hexavalent 5f-in-core LPPs [30] were performed [52]. Corresponding to the calibration studies on AnF_3 (An=Ac–Lr) [19, 20] using the trivalent 5f-in-core LPPs, calculations using these LPPs in connection with CPPs [54] were carried out [52]. Since for the actinide fluorides only a few experimental and AE data

---
[2]The LPP HF calculation for BkF_4 was performed by X. Cao.

3.1 5F-IN-CORE PSEUDOPOTENTIALS FOR ACTINIDES

are available, 5f-in-valence SPP [15] calculations[3] were performed [52] as well. For F Dunning's aug-cc-pVQZ (augmented correlation-consistent polarized VQZ) basis set [62, 63] was applied and for An (7s6p5d2f1g)/[6s5p4d2f1g] and (14s13p10d8f6g)/ [6s6p5d4f3g] [21] valence basis sets[4] were used for LPP HF and SPP state-averaged MCSCF calculations, respectively. The state-averaging was necessary to avoid symmetry-breaking at the orbital level, since MOLPRO [52] is limited to the D_{2h} point group and subgroups. The geometries were completely optimized imposing C_{2v}, C_{3v}, T_d, C_{4v}, and O_h symmetry for AnF_2, AnF_3, AnF_4, AnF_5, and AnF_6, respectively. In the case of ThF_4, UF_4, PaF_5, UF_5, and UF_6-PuF_6 also LPP CCSD(T) calculations were performed, because for these compounds experimental [64–68] or AE [69] results are available. Moreover, for AmF_2, NoF_2, and all AnF_3 LPP CCSD(T) calculations with and without using CPPs were carried out. Additionally, for AmF_2, NoF_2, and LrF_3 SPP CCSD(T) calculations were performed. In the CCSD(T) calculations the F 1s orbitals were frozen.

The An–F bond energies were calculated by $E_{bond} = [E(An) + n \times E(F) - E(AnF_n)]/n$ (with $n=2$, 4–6 for AnF_2, AnF_4–AnF_6), where the actinide atom was assumed to be in the lowest valence substate, i.e. $5f^{n+1}7s^2$, $5f^{n-1}6d^27s^2$, $5f^{n-2}6d^37s^2$, and $5f^{n-3}6d^47s^2$ for AnF_2, AnF_4–AnF_6, respectively. At this point one might ask how to calculate a binding energy with respect to the experimentally observed ground states of the actinides, e.g. at the correlated level. The best way is to follow the strategy proposed for the lanthanide LPPs almost two decades ago [70]. First, one should calculate the binding energy with respect to the actinide atom in its lowest valence substate. Then the energy difference to the experimentally observed ground state, possibly belonging to a different configuration, can be determined, e.g. at the AE WB [55] or DHF level [44], and corrected by electron correlation contributions to the energy difference (between the lowest levels) taken from experiment [71]. However, in contrast to di-, tri-, and tetravalent PPs for penta- and hexavalent PPs an energy correction using experimental energy differences is not possible, since for the $6d^37s^2$ and $6d^47s^2$ valence subconfigurations no experimental data are available [71]. If desired, correlation contributions can of course be obtained by 5f-in-valence SPP or AE calculations. Tables C.2 and C.3, C.4 and C.5, and C.6 summarize AE WB, AE DHF, and experimental corrections for the di-, tri-, tetra-, penta-, and hexavalent LPPs, respectively.

---

[3] The SPP calculations for AnF_4 (An=Th–Cf) were performed by X. Cao.
[4] For AnF_3 (An=Ac–Lr) the LPP and SPP calculations were carried out using (7s6p5d2f1g)/ [5s4p3d2f1g] and (14s13p10d8f6g)/[10s9p5d4f3g] valence basis sets, respectively.

For AnF$_3$ ionic binding energies $\Delta E_{ion} = E(An^{3+}) + 3 \times E(F^-) - E(AnF_3)$ were calculated.

3.1.5.2 Actinide Difluorides

The HF calculations for AnF$_2$ (An=Pu–No) using LPPs with and without CPPs will be compared to corresponding SPP calculations. The comparison is reasonable for the late actinides, but critical for the lighter actinides due to mixing of 5f with 7s as well as 6d orbitals in the SPP calculations. This can best be seen from the calculated bond angles, which show much larger discrepancies for Pu–Cm than for Bk–No (cf. Fig. 3.4). Hence, all results will be compared separately for the elements Pu–Cm and Bk–No, respectively. The results of a Mulliken population analysis and those for structures as well as binding energies of the LPP, LPP+CPP, and SPP calculations are listed in Table 3.4 and 3.5, respectively.

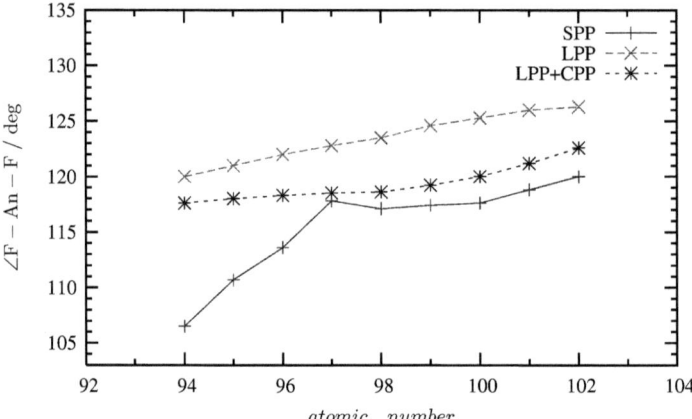

Figure 3.4: Actinide–fluorine bond angles ∠F–An–F (in deg) for AnF$_2$ (An=Pu–No) from LPP HF calculations with and without using CPPs as well as from SPP state-averaged MCSCF calculations.

Mulliken Orbital Populations Table 3.4 shows the Mulliken orbital populations obtained by LPP HF and SPP state-averaged MCSCF calculations, respectively. As one can see the bonding of AnF$_2$ is of polar covalent nature, i.e. the two binding electron

3.1 5F-IN-CORE PSEUDOPOTENTIALS FOR ACTINIDES

Table 3.4: Mulliken 6s/7s, 6p/7p, 6d, and 5f orbital populations and atomic charges (Q) on An in AnF$_2$ (An=Pu–No) from LPP HF and SPP state-averaged MCSCF calculations. A $6s^2 6p^6 7s^2$ ground state valence subconfiguration is considered for An.

An	s LPP	s SPP	p LPP	p SPP	d LPP	d SPP	f LPP[a]	f SPP	Q LPP	Q SPP
Pu	2.00	2.16	6.00	6.00	0.21	0.37	0.02	5.90	1.78	1.57
Am	2.00	2.12	6.00	6.00	0.21	0.32	0.01	6.97	1.78	1.58
Cm	2.00	2.12	5.99	6.00	0.20	0.30	0.01	8.00	1.79	1.59
Bk	2.00	2.12	5.99	6.01	0.20	0.28	0.01	9.01	1.79	1.59
Cf	2.00	2.13	5.99	6.01	0.19	0.28	0.01	10.01	1.80	1.57
Es	2.01	2.14	5.99	6.02	0.19	0.27	0.01	11.01	1.80	1.56
Fm	2.01	2.14	5.99	6.02	0.19	0.27	0.01	12.02	1.80	1.54
Md	2.01	2.15	5.99	6.02	0.18	0.26	0.01	13.01	1.80	1.56
No	2.02	2.15	5.99	6.02	0.18	0.25	0.01	14.01	1.80	1.57

[a] 6–14 electrons in the 5f shell are attributed to the LPP core for Pu–No, respectively.

pairs are dragged more close to the fluorine ends of the bonds. For the LPP calculations this results in charge separations of up to 0.90 electrons per bond and a total atomic charge of up to 1.80 units on the actinide. Whereas the s, p, and f occupation numbers on the actinides are nearly integral, those of the d shells are not and point to some covalent contributions. The SPP f orbital occupations show that there is almost no 5f orbital participation in the bonding of AnF$_2$ with An=Cm–No, since the SPP 5f populations differ at most by 0.02 electrons from the integral LPP 5f occupations. However, for PuF$_2$ and AmF$_2$ the SPP calculations do not yield 5f occupations corresponding to a divalent actinide, i.e. the 5f populations are 0.10/0.03 electrons below the integral number of 5f electrons for Pu/Am. This is due to the stronger mixing between 5f and 7s as well as 6d orbitals for these lighter actinides, where the 5f orbitals are still relatively diffuse. The mixing of 5f with 6d orbitals, which can also be seen as a configurational mixing of $5f^{n+1}$ and $5f^n 6d^1$, decreases from Pu to No, since the 6d orbitals are destabilized due to the indirect relativistic effect, and are thus less occupied (6d AE WB orbital energies for $5f^{n+1} 6s^2 6p^6 6d^1 7s^1$: $-3.123/-2.311$ eV for Pu/No; 6d occupation for SPP: 0.37/0.25 for Pu/No). Therefore the LPP results are expected to become better with increasing nuclear charge and to be less accurate for the lighter actinides Pu–Am, where the 5f occupation falls below the assumed integral value corresponding to a divalent actinide.

Actinide Difluoride Structure For all AnF$_2$ (An=Pu–No) non-linear structures were obtained by both LPP and SPP calculations. In the case of the LPP calculations for increasing nuclear charge of the actinide the bond lengths decrease almost linearly (correlation coefficient 0.999) by 0.08 Å and the bond angles increase by 6°. These variations are due to the actinide contraction and the increasing repulsion between the fluorine atoms, respectively. Similar trends are also observed for the SPP results, i.e. bond lengths decrease and bond angles increase by 0.03 Å and 13°, respectively, however here two irregularities appear. On the one hand the bond length increases instead of decreases from Pu to Am by 0.01 Å, and on the other hand the bond angles between Pu and Bk increase on average by 4°, while those between Bk and No grow only by about 0.4°. The reason for both is a mixing of 5f with 7s as well as 6d orbitals, which becomes more significant with decreasing nuclear charge, and as already mentioned above limits the applicability of the 5f-in-core approach.

A comparison of An–F bond lengths calculated using LPPs and SPPs demonstrates that the newly developed LPPs yield quite accurate results for all actinides considered. The bond lengths are on average by 0.042 (0.020) Å and 1.9% (0.9%) too long for Pu–Cm (Bk–No). The actinide–fluorine bond angles ∠F–An–F, which are also overestimated by using LPPs, show clearly larger deviations. The m.a.e. and m.r.e. for Pu–Cm (Bk–No) amount to 10.7° (6.6°) and 10% (6%), respectively. The largest deviations for both bond lengths and angles occur for Pu (0.059 Å; 13.5°) and Am (0.037 Å; 10.3°), because here the SPP 5f occupation is smaller than the integral value modeled by the LPP core (cf. Table 3.4).

The use of LPPs in connection with CPPs gives about 0.036 Å and 4.2° smaller bond lengths and angles, respectively. Since pure LPP calculations overestimate the SPP An–F bond lengths by ca. 0.027 Å, they are underestimated by ca. 0.015 Å using LPPs in combination with CPPs (m.a.e. for Pu–No). Considering the deviations in bond lengths for Pu–Cm and Bk–No separately, one finds a clear improvement by using CPPs in the case of Pu–Cm, i.e. the mean deviation related to the SPP data decreases by 0.029 Å (m.a.e.: LPP/LPP+CPP 0.042/0.013 Å). For Bk–No, however, this deviation remains almost constant, i.e. the improvement in the m.a.e. amounts to 0.005 Å (m.a.e.: LPP/LPP+CPP 0.020/0.015 Å). In the case of the bond angles the decrease by using CPPs reduces the errors of the LPP calculations for all actinides considered, i.e. the m.a.e. for Pu–No drops from 8.0 to 3.9°.

3.1 5F-IN-CORE PSEUDOPOTENTIALS FOR ACTINIDES

Table 3.5: An–F bond lengths R_e (in Å) and angles $\angle$F–An–F (in deg) as well as bond energies E_{bond} (in eV) for AnF$_2$ (An=Pu–No) from LPP HF calculations with and without using CPPs as well as from SPP state-averaged MCSCF calculations. For AmF$_2$ and NoF$_2$ LPP CCSD(T) results with and without using CPPs as well as SPP CCSD(T) results are given in italics.

An	R_e			$\angle$F–An–F			E_{bond}		
	LPP	CPP[a]	SPP	LPP	CPP[a]	SPP	LPP	CPP[a]	SPP
Pu	2.212	2.179	2.152	120.0	117.6	106.5	3.764	3.644	4.057
Am	2.200	2.164	2.163	121.0	118.0	110.7	3.715	3.605	3.907
	2.182	*2.157*	*2.089*	*116.7*	*114.6*	*104.3*	*5.188*	*5.209*	*5.517*
Cm	2.189	2.150	2.161	122.0	118.3	113.6	3.667	3.571	3.836
Bk	2.178	2.137	2.155	122.8	118.5	117.8	3.617	3.540	3.764
Cf	2.168	2.127	2.144	123.5	118.6	117.1	3.564	3.507	3.705
Es	2.156	2.116	2.134	124.6	119.2	117.4	3.517	3.476	3.611
Fm	2.146	2.110	2.122	125.3	120.0	117.6	3.464	3.434	3.540
Md	2.136	2.106	2.120	126.0	121.2	118.8	3.409	3.380	3.468
No	2.128	2.105	2.118	126.3	122.6	120.0	3.347	3.312	3.375
	2.114	*2.100*	*2.057*	*121.2*	*118.9*	*112.5*	*4.790*	*4.833*	*5.043*

[a]LPP calculations using CPPs.

Actinide–fluorine Bond Energy   The An–F bond energy of AnF$_2$ decreases by 0.42 and 0.68 eV with increasing nuclear charge for LPP and SPP calculations, respectively. This is related to the decreasing An–F bond length, which is accompanied by an increasing F–F repulsion.

The differences in the An–F bond energies between LPP and SPP calculations for the lighter actinides are obviously larger than those for the heavier actinides, i.e. the m.a.e. (m.r.e.) for Pu–Cm and Bk–No are 0.22 (5.5%) and 0.09 eV (2.5%), respectively. This is most likely due to a mixing of valence 5f with mainly valence 7s and 6d orbitals in the SPP state-averaged MCSCF calculations. Analogous to the bond lengths and angles the largest errors occur for Pu (0.29 eV, 7.2%) and Am (0.19 eV, 4.9%), where the SPP 5f occupations are smaller than the assumed integral LPP 5f occupations (cf. Table 3.4).

The application of CPPs causes a mean decrease in bond energy of 0.066 eV compared to pure LPP calculations. Since the bond energy is already underestimated by using LPPs without CPPs, the deviations from SPP calculations become larger by using CPPs. The m.a.e. (m.r.e.) amount to 0.33 eV (8.3%) and 0.14 eV (3.7%) for Pu–Cm and Bk–No, respectively. The strong energy decrease due to CPPs can be explained, if one thinks of an ionic binding energy

$$E_{ion} = -IP_1(An) - IP_2(An) + 2 \times EA(F) + \text{ionic interaction} . \quad (3.4)$$

Here $IP_i(An)$ with $i=1, 2$ are the first and second IPs of the actinide, respectively, and EA(F) is the electron affinity of fluorine. The use of CPPs increases the IPs, because the actinide atom or ion is stabilized via the included correlation, and thus the bond energy is reduced. The first and second IPs of Pu–No from LPP state-averaged MCSCF calculations with and without using CPPs in comparison to SPP state-averaged MCSCF data [21] are listed in Table 3.6. The IPs of Cm are omitted, because its ionizations do not take place between divalent oxidation states, i.e. the Cm atom and Cm^+ cation do not have a $5f^{n+1}$, but a $5f^n$ occupation. As one can see IP_1 and IP_2 are increased by about 0.32 and 0.50 eV, respectively, if LPP+CPP instead of LPP state-averaged MCSCF calculations are performed. The pure LPP IPs deviate only on average by 0.09/0.12 eV from the SPP values for IP_1/IP_2, however, using CPPs the IPs are overestimated by about 0.39/0.38 eV. The reason for the worse IP results using CPPs, which also explains the larger deviations of An–F bond energies, might be the inclusion of dynamic correlation. Since the CPPs are adjusted to CCSD(T) reference data, they account for both static (polarization at the HF level) and dynamic (core–valence correlation) polarization of the PP core, even if they are applied in HF or state-averaged MCSCF calculations. The SPP state-averaged MCSCF calculations, however, do not include any correlation effects. Thus, the IPs calculated by using CPPs become too large and the corresponding bond energies are too small. However, at the correlated level the experimental and SPP ACPF IPs are always underestimated by the LPP CCSD(T) results and the m.a.e. amount to 0.26 and 0.21 eV for IP_1 and IP_2, respectively (m.a.e. for Pu–No except for Cm, cf. Table C.1). Therefore the application of CPPs yields improved results, i.e. the m.a.e. are reduced to 0.14 and 0.04 eV for IP_1 and IP_2, respectively.

A slight improvement due to CPPs is also found in CCSD(T) calculations for AmF_2 and NoF_2 (cf. Table 3.5). Here, the deviations from the SPP CCSD(T) bond energies amount to 0.33/0.31 and 0.25/0.21 eV for LPP/LPP+CPP calculations of AmF_2 and NoF_2, respectively. Furthermore, the LPP+CPP bond lengths and bond angles are also in better agreement with the SPP data than the pure LPP results (differences between SPP and LPP/LPP+CPP: AmF_2 ΔR_e=0.093/0.068 Å, $\Delta\angle$=12.4/10.3°; NoF_2 ΔR_e=0.057/0.043 Å, $\Delta\angle$=8.7/6.4°).

3.1 5F-IN-CORE PSEUDOPOTENTIALS FOR ACTINIDES

Table 3.6: First and second IPs (in eV) of the divalent actinides Pu–No (except for Cm) from LPP state-averaged MCSCF calculations with and without using CPPs in comparison to SPP state-averaged MCSCF data [21].[a]

An	IP_1^b			IP_2^b		
	LPP	CPP[c]	SPP	LPP	CPP[c]	SPP
Pu	4.91	5.22	4.73	10.65	11.18	10.84
Am	4.97	5.30	4.77	10.80	11.36	11.02
Bk	5.11	5.46	4.99	11.10	11.66	11.25
Cf	5.18	5.53	5.09	11.24	11.79	11.34
Es	5.25	5.59	5.21	11.40	11.92	11.50
Fm	5.32	5.64	5.34	11.54	12.03	11.66
Md	5.39	5.69	5.41	11.69	12.12	11.75
No	5.46	5.71	5.51	11.84	12.18	11.88

[a] Basis sets: LPP (7s6p5d2f1g)/[6s5p4d2f1g]; SPP (14s13p10d8f6g)/[6s6p5d4f3g].
[b] Initial and final states: IP_1: $5f^{n+1}7s^2 \to 5f^{n+1}7s^1$; IP_2: $5f^{n+1}7s^1 \to 5f^{n+1}7s^0$.
[c] LPP calculations using CPPs.

Aside from the limited validity of the 5f-in-core approach for PuF_2 and AmF_2, the deviations can be explained by the larger BSSE of the SPP compared to the LPP/LPP+CPP calculations at the CCSD(T) level. Using the counterpoise correction the SPP bond energies are reduced from 5.517/5.043 eV to 5.319/4.777 eV corresponding to a BSSE of 0.198/0.266 eV for AmF_2/NoF_2. These are reasonable amounts for the BSSE, since the (14s13p10d8f6g)/[6s6p5d4f3g] basis sets [21] recover only about 80% of the atomic CCSD(T) correlation energy. In the case of the LPP CCSD(T) calculations the counterpoise correction yields by 0.041/0.039 eV smaller bond energies, i.e. the energies are reduced from 5.188/4.790 eV to 5.147/4.751 eV for AmF_2/NoF_2. Thus, the BSSE using LPPs are clearly smaller than that using SPPs, which constitutes an enormous advantage compared to the SPP calculations. Taking the BSSE into account the deviations in bond energies related to the SPP data are reduced to 0.17/0.15 eV and 0.026/0.017 eV for LPP/LPP+CPP calculations of AmF_2 and NoF_2, respectively. Hence, at the correlated level the LPP bond energies with and without using CPPs are in good agreement with the reference data, and the use of CPPs shows an improvement of the results.

3.1.5.3 Actinide Tetrafluorides

The HF and CCSD(T) calculations for AnF_4 (An=Th–Cf) using LPPs with and without CPPs will be compared to SPP and experimental data [64, 65]. The Mulliken orbital population analysis will not be discussed in detail, because it leads to similar conclusions as for AnF_2. However, the 5f orbital populations will be given together with the other results as well as the available experimental data in Table 3.7.

Actinide Tetrafluoride Structure The An–F bond lengths decrease almost linearly (LPP correlation coefficient 0.995) with increasing nuclear charge, whereby the decrease from ThF_4 to CfF_4 amounts to 0.09 and 0.11 Å for LPP HF and SPP state-averaged MCSCF calculations, respectively. Since for AnF_4 as well as for AnF_2 nine actinide elements, i.e. Th–Cf respectively Pu–No, are considered, one can compare the actinide contraction for these compounds. In the case of AnF_4 the contraction is somewhat larger than that for AnF_2 (AnF_2 contraction: 0.08/0.03 Å for LPP/SPP). The reason for this is that the An–F bond in AnF_2 is more 'rigid' [72] as can be seen from a comparison of the force constants, e.g., for LPP HF calculations the force constants are 0.06427 and 0.05843 a.u. for PuF_2 and PuF_4, respectively.

The LPP HF results are in good agreement with the SPP reference data, i.e. the An–F distances determined using LPPs are at most 0.023 Å (1.1%) too long. The m.a.e. and m.r.e. amount to 0.018 Å and 0.9%, respectively. For CmF_4 the LPP HF result differs by 0.022 Å (1.1%) from the AE DHF bond length 2.022 Å [73] determined by using (28s28p19d13f2g) and (13s9p3d) basis sets for Cm and F, respectively. Compared to experimental (Th, U) and estimated (Np–Am) values the LPP HF calculations yield also satisfactory results, i.e. the m.a.e. (m.r.e.) amounts to 0.034 Å (1.7%). The obtained An–F bond lengths (An=Th, U–Am) are also in good agreement with those determined by an interionic force model (Th 2.140, U 2.055, Np 2.042, Pu 2.029, Am 2.017 Å) [74], i.e. the bond lengths deviate on average by 0.039 Å (1.9%). So the 5f-in-core approximation holds quite good for the tetravalent PPs, although the calculated SPP 5f occupations are about 0.22 electrons larger than the integral LPP occupations. The reason why this approximation still works is the f-part of the LPP, which sufficiently accounts for the 5f participation in chemical bonding, i.e. the LPP 5f occupations attain values up to 0.27 electrons, and thus the differences between the LPP and SPP 5f occupations only amount to ca. 0.07 electrons.

3.1 5F-IN-CORE PSEUDOPOTENTIALS FOR ACTINIDES

Table 3.7: An–F bond lengths R_e (in Å), bond energies E_{bond} (in eV), and f orbital occupations for AnF$_4$ (An=Th–Cf) from LPP HF calculations with and without using CPPs in comparison to experimental/estimated data [64, 65] and SPP state-averaged MCSCF calculations. For ThF$_4$ and UF$_4$ LPP CCSD(T) results with and without using CPPs are given in italics.

	R_e				E_{bond}			f occ.	
An	LPP	CPP[a]	SPP	Exp.[b]	LPP	CPP[a]	SPP	LPP[c]	SPP
Th	2.107	2.101	2.115	2.124	5.617	5.630	5.571	0.27	0.28
	2.101	*2.097*			*7.117*	*7.152*			
Pa	2.104	2.098	2.092		5.422	5.426	5.451	0.20	1.23
U	2.094	2.088	2.072	2.059	5.311	5.310	5.379	0.16	2.22
	2.091	*2.088*			*6.813*	*6.839*			
Np	2.082	2.075	2.059	2.04	5.240	5.237	5.254	0.15	3.22
Pu	2.070	2.063	2.047	2.03	5.188	5.183	5.146	0.13	4.22
Am	2.057	2.050	2.035	2.02	5.152	5.147	5.057	0.12	5.22
Cm	2.044	2.037	2.026		5.131	5.124	4.980	0.12	6.21
Bk	2.031	2.024	2.017		5.118	5.112	4.880	0.11	7.19
Cf	2.020	2.013	2.001		5.106	5.099	5.037	0.11	8.19

[a] LPP calculations using CPPs.
[b] For Np–Am the values are estimated.
[c] 0–8 electrons in the 5f shell are attributed to the LPP core for Th–Cf, respectively.

The use of LPPs in combination with CPPs yields about 0.007 Å smaller bond lengths compared to pure LPP HF calculations. Therefore the deviations from SPP calculations are reduced by about 30% compared to those using pure LPPs, i.e. the m.a.e. (m.r.e.) decreases from 0.018 (0.9%) to 0.012 Å (0.6%). The comparison to experimental and estimated data shows just a slight improvement, if CPPs are used, i.e. the m.a.e. (m.r.e.) decreases from 0.034 (1.7%) to 0.030 Å (1.5%).

The introduction of correlation via CCSD(T) shortens the An–F distances by 0.006 and 0.003 Å for ThF$_4$ and UF$_4$, respectively. The differences between LPP CCSD(T) calculations and experimental data amount to $-0.023/+0.032$ Å for ThF$_4$/UF$_4$, and are thus slightly larger respectively smaller than those between LPP HF calculations and the experiment ($-0.017/+0.035$ Å for ThF$_4$/UF$_4$).

The use of CPPs at the CCSD(T) level has only a small effect on the An–F bond length, i.e. the Th–F and U–F bond lengths are shortened by 0.004 and 0.003 Å, respectively. Therefore the deviation to the experiment also changes only slightly, i.e. it becomes larger respectively smaller for ThF$_4$ and UF$_4$ ($-0.023/-0.027$ and $+0.032/+0.029$ Å for LPP/LPP+CPP of ThF$_4$ and UF$_4$).

Actinide–fluorine Bond Energy The An–F bond energies decrease by 0.51 and 0.53 eV from ThF$_4$ to CfF$_4$ for LPP and SPP calculations, respectively. This is due to the increasing F–F repulsion with decreasing An–F distances as it is the case for AnF$_2$. While the LPP HF bond energies decrease smoothly, the SPP data show a minimum for BkF$_4$, i.e. for the half-filled 5f shell.

The LPP HF bond energies deviate at most by 0.15 eV (3.0%) from SPP reference data except for BkF$_4$, for which the difference is 0.24 eV (4.9%). However, this deviation is reduced to 0.049 eV (0.7%), if LPP and SPP CCSD(T) single-point calculations on the HF optimized BkF$_4$ structures are compared (CCSD(T) results: 6.604/6.653 eV for LPP/SPP; frozen orbitals: F 1s for LPP and F 1s, Bk 5s, 5p, 5d for SPP). Taking the BSSE into account this deviation is even further reduced to 0.037 eV (counterpoise corrected CCSD(T) results: 6.507/6.470 eV for LPP/SPP). To investigate the correlation effects single-point CCSD(T) calculations were considered to be sufficient, since the AnF$_4$ structures are only slightly affected by using CCSD(T) instead of HF, i.e. the An–F bond lengths decrease by at most 0.006 Å. For the other elements (Th–Cm, Cf) the m.a.e. (m.r.e.) at the HF level amounts to 0.064 eV (1.2%) and the largest deviations occur for Am (0.095 eV) and Cm (0.15 eV), where the differences between the LPP and SPP 5f occupations achieve their maximum (0.10/0.09 electrons for Am/Cm). The application of CPPs at the HF level affects the bond energies only very slightly and the deviations compared to the SPP data remain almost constant, i.e. without BkF$_4$ the m.a.e. (m.r.e.) is 0.063 (1.2%). The change from LPP HF to LPP CCSD(T) calculations results in a strong increase of the bond energies by 1.50 eV. The use of LPPs in connection with CPPs at the CCSD(T) level causes for both ThF$_4$ and UF$_4$ only a small increase in bond energy by 0.035 and 0.026 eV, respectively.

3.1.5.4 Actinide Pentafluorides

The LPP HF and CCSD(T) calculations for AnF$_5$ (An=Pa–Am) will be compared to SPP state-averaged MCSCF calculations as well as to experimental [66] and computational [69, 75, 76] data from the literature. The results for bond lengths, angles, and energies as well as the 5f orbital populations are listed in Table 3.8.

Actinide Pentafluoride Structure While in 1977 the infrared spectrum of UF$_5$ [66] indicated a C_{4v} symmetry, later computational studies [75] including relativistic effects (also SO coupling) showed the D_{3h} geometry to be 1 kcal mol^{-1} lower than the C_{4v}

3.1 5F-IN-CORE PSEUDOPOTENTIALS FOR ACTINIDES

one. This finding is not contradictory to the experimental result, since in the photogeneration of UF_5 from UF_6 one has an internal energy excess of more than 1 kcal mol^{-1}. The geometry optimizations were performed imposing C_{4v} symmetry, so that the LPP results can also be compared to experimental values.

Due to the actinide contraction the An–F bond lengths calculated by using LPP HF and SPP state-averaged MCSCF decrease continuously with increasing nuclear charge. The decrease of axial bond lengths R_{ax} is slightly larger than that of equatorial bond lengths R_{eq}, because the axial ligand experiences a lower ligand–ligand repulsion than the equatorial ligands (LPP: ΔR_{ax}=0.05, ΔR_{eq}=0.03; SPP: ΔR_{ax}=0.07, ΔR_{eq}=0.05 Å). The F_{ax}–An–F_{eq} bond angles stay almost constant along the actinide row, i.e. the deviation between angles of different actinides amount at most to 0.9 and 1.8° for LPPs and SPPs, respectively.

Table 3.8: An–F bond lengths R_{ax} and R_{eq} (in Å) and angles $\angle F_{ax}$–An–F_{eq} (in deg), bond energies E_{bond} (in eV), and f orbital populations for AnF_5 (An=Pa–Am) from LPP HF and SPP state-averaged MCSCF calculations. For PaF_5 and UF_5 LPP CCSD(T) and AE DFT [69] as well as experimental [66] data are given in italics.

An	R_{ax}		R_{eq}		$\angle$		E_{bond}		f occ.	
	LPP	SPP[a]	LPP	SPP[a]	LPP	SPP[a]	LPP	SPP	LPP[b]	SPP
Pa	2.034	2.047	2.027	2.040	106.8	105.6	5.556	5.504	0.56	0.56
	2.035	*2.061*	*2.023*	*2.060*	*106.4*	*100.3*	*7.086*			
U	2.028	2.026	2.028	2.022	107.5	105.7	5.263	5.386	0.42	1.55
	2.032	*2.00*	*2.028*	*2.02*	*107.4*	*101*	*6.759*			
Np	2.017	2.008	2.021	2.007	107.7	103.9	5.099	5.255	0.36	2.60
Pu	2.002	1.996	2.009	1.996	107.7	104.7	5.003	5.092	0.33	3.61
Am	1.988	1.982	1.998	1.987	107.6	104.5	4.933	4.962	0.31	4.63

[a] Given in italics: PaF_5: AE DFT/BP86 results using ZORA and pVTZ basis sets; UF_5: experimental values.
[b] 0–4 electrons in the 5f shell are attributed to the LPP core for Pa–Am, respectively.

The An–F bond lengths R_{ax} and R_{eq} from LPP HF calculations are in good agreement with the SPP reference data, i.e. the m.a.e. (m.r.e.) amount to 0.007 (0.4%) and 0.011 Å (0.5%) for R_{ax} and R_{eq}, respectively. The maximum error for both bond lengths is 0.013 Å (0.7%). The deviations between LPP and SPP bond angles are slightly larger, i.e. the m.a.e. (m.r.e.) and the maximum error amount to 2.6 (2.5%) and 3.8° (3.7%), respectively. Furthermore, the LPP HF structure for UF_5 is comparable to that of a former HF calculation, where a Cowan–Griffin ECP for U and VDZ basis sets were used (ECP HF: R_{ax}=2.00, R_{eq}=2.00 Å, ∠=100°) [75]. The deviations in bond lengths and angles are 0.03 Å (1.4%) and 8° (8.0%), respectively.

If correlation is included via CCSD(T), the molecular structures of PaF$_5$ and UF$_5$ change only slightly by at most 0.004 Å and 0.4°. For PaF$_5$ the deviations between LPP CCSD(T) results and AE DFT/BP86 calculations using the zero-order regular approximation (ZORA) and pVTZ basis sets [69] amount to 0.026, 0.037 Å, and 6.1° for R_{ax}, R_{eq}, and ∠, respectively. For UF$_5$ the differences between LPP CCSD(T) and experimental [66] results are 0.03, 0.01 Å, and 6° for R_{ax}, R_{eq}, and ∠, respectively. Moreover, the comparison to the UF$_5$ structure calculated by SPP DFT/PBE0 using pVDZ basis sets [76] gives deviations of just 0.018, 0.013 Å, and 9.1° for R_{ax}, R_{eq}, and ∠, respectively (SPP DFT/PBE0: R_{ax}=2.014, R_{eq}=2.015 Å, ∠=98.3°). Thus, the LPP CCSD(T) results are also in good agreement with corresponding reference data and confirm the reliability of the pentavalent LPPs.

These good results can be explained by the 5f occupations. The SPP 5f occupations vary on average by 0.59 and at most by 0.63 electrons from the assumed LPP $5f^{n-2}$ occupations, which demonstrates that the 5f orbitals participate to some extent in the An–F bonding. However, the 5f-in-core approach still yields reasonable results, since the differences between LPP and SPP 5f occupations amount on average only to 0.19 and at most to 0.32 electrons, because the f-part of the LPP allows for some 5f occupation in addition to the integral $5f^{n-2}$ assumption.

Actinide–fluorine Bond Energy The An–F bond energy of AnF$_5$ decreases by 0.62 and 0.54 eV with increasing nuclear charge for LPP HF and SPP state-averaged MCSCF calculations, respectively. This is related to the increasing F–F repulsion, which is due to the decreasing An–F bond length.

The LPP and SPP An–F bond energies are in good agreement, i.e. the m.a.e. (m.r.e.) amounts to 0.090 eV (1.7%) and the maximum error, which occurs for Np, is 0.16 eV (3.0%). As expected the inclusion of electron correlation via CCSD(T) clearly increases the An–F bond energies by ca. 1.5 eV.

3.1.5.5 Actinide Hexafluorides

The LPP HF and CCSD(T) calculations for AnF$_6$ (An=U–Am) will be compared to SPP state-averaged MCSCF results and experimental [67, 68] as well as computational [13, 77, 78] data from the literature. The results for bond lengths, bond energies, and

5f orbital populations are listed in Table 3.9.

Actinide Hexafluoride Structure   The An–F bond lengths calculated by using LPPs at the HF level increase from UF_6 to NpF_6 by 0.007 Å and decrease from NpF_6 to AmF_6 by 0.018 Å. The SPP state-averaged MCSCF bond lengths, however, decrease smoothly with increasing nuclear charge by 0.034 Å. While the reason for the decrease is the well-known actinide contraction, the increase from UF_6 to NpF_6 is possibly due to a shortcoming of the LPP method, because for R_{eq} of AnF_5 an analogous, but clearly smaller, increase from PaF_5 to UF_5 by 0.001 Å is obtained (cf. Table 3.8). Since the U–F bond length is only by ca. 0.02 Å smaller than expected, this LPP shortcoming is still acceptable.

Table 3.9: An–F bond lengths R_e (in Å), bond energies E_{bond} (in eV), and f orbital populations for AnF_6 (An=U–Am) from LPP HF and SPP state-averaged MCSCF calculations. For UF_6, NpF_6, and PuF_6 LPP CCSD(T) and experimental [67] results are given, too.

	R_e				E_{bond}			f occ.	
An	LPP[a]	SPP	CC[b]	Exp.[c]	LPP[a]	SPP	CC[b]	LPP[a,d]	SPP
U	1.976	1.975	1.978	1.996(8)	5.355	5.646	6.931	0.96	1.16
Np	1.983	1.966	1.988	1.981(8)	4.813	5.499	6.360	0.67	2.28
Pu	1.980	1.949	1.989	1.971(10)	4.542	5.439	6.093	0.56	3.46
Am	1.965	1.941			4.479	5.279		0.59	4.52

[a]LPP HF results.
[b]LPP CCSD(T) results.
[c]For UF_6 also another experimental value is available: R_e=1.999(3) Å [68].
[d]0–3 electrons in the 5f shell are attributed to the LPP core for U–Am, respectively.

The LPP HF results are in good agreement with the SPP reference data, i.e. the LPP An–F distances are at most by 0.031 Å (1.6%) too long and the m.a.e. (m.r.e.) amounts to 0.018 Å (0.9%). For UF_6, NpF_6, and PuF_6 the comparison of the LPP HF bond lengths to those of HF calculations [13], where Cowan–Griffin ECPs for An and pVDZ basis sets were used, shows also satisfactory results, i.e. the maximum error amounts to 0.037 Å (1.9%) (ECP HF: UF_6 R_e=1.984, NpF_6 R_e=1.972, PuF_6 R_e=1.943 Å). Moreover, the obtained U–F bond length is also in good agreement with that determined by a SPP HF calculation [78] using an aug-pVDZ basis set for F, i.e. the bond lengths deviate by 0.009 Å (0.5%) (SPP HF: R_e=1.985 Å). Thus, in the case of the An–F bond lengths the hexavalent 5f-in-core approximation still holds, although the calculated SPP 5f occupations are about 1.35 electrons larger than the integral LPP occupations and even the differences between the LPP and SPP 5f occupations amount to ca. 0.66 electrons.

Analogous to the AnF_5 results the introduction of correlation via CCSD(T) increases the An–F bond lengths only slightly by at most 0.009 Å. The differences between LPP CCSD(T) and experimental [67] data are at most 0.018 Å (0.9%) and the m.a.e. (m.r.e.) amounts to 0.014 Å (0.7%). Compared to DFT data from the literature [13, 77, 78] the LPP CCSD(T) bond lengths deviate at most by 0.036 Å (1.8%) (ECP DFT/B3LYP [13]: UF_6 R_e=2.014, NpF_6 R_e=2.013, PuF_6 R_e=1.985; SPP DFT/B3LYP [77]: UF_6 R_e=2.007, NpF_6 R_e=1.991, PuF_6 R_e=1.977; SPP DFT/PBE0 [78]: UF_6 R_e=1.994 Å). Therefore the correlated calculations of AnF_6 (An=U–Pu) confirm the good performance of the hexavalent 5f-in-core PPs for the An–F bond lengths.

Actinide–fluorine Bond Energy The An–F bond energies decrease continuously with increasing nuclear charge by 0.88 and 0.37 eV for LPP and SPP calculations, respectively. This is due to the increasing F–F repulsion with decreasing An–F distances as it is the case for AnF_5.

In contrast to the good agreement for the An–F bond lengths, the LPP HF bond energies of AnF_6 deviate considerably from the SPP state-averaged MCSCF data, i.e. the m.a.e. (m.r.e.) and the maximum difference are 0.67 (12.3%) and 0.90 eV (16.5%), respectively. These significant discrepancies are in line with the high differences between LPP and SPP 5f occupations of up to 0.93 electrons. For UF_6, however, where these 5f occupations differ only by 0.20 electrons, the An–F bond energy is still reasonable, i.e. it deviates by 0.29 eV (5.1%). Thus, the 5f-in-core approximation seems to reach its limitations for the hexavalent oxidation state except for U, which corresponds to $5f^0$. Therefore the hexavalent 5f-in-core LPPs for Np–Am should only be used for preoptimizing structures prior to more rigorous studies including the 5f shell explicitly.

3.1.5.6 Actinide Trifluorides

The HF calculations for AnF_3 (An=Ac–Lr) using LPPs in connection with CPPs will be compared to those using pure LPPs respectively SPPs [20] (cf. Table 3.10). Furthermore, CCSD(T) calculations using LPPs with and without CPPs will be compared to DFT, MP2, and complete active space second-order perturbation theory (CASPT2)

3.1 5F-IN-CORE PSEUDOPOTENTIALS FOR ACTINIDES

data from the literature [76, 79, 80] (cf. Table 3.11).

Since for Th and Pa the trivalent oxidation state is not preferred (Th) or even not stable (Pa) in aqueous solution (cf. Table 1.1) [1], the trivalent subconfiguration $5f^n$ mixes strongly with the corresponding energetically low-lying tetravalent subconfiguration $5f^{n-1}6d^1$ yielding significantly smaller SPP 5f occupations than assumed for the LPP core [20] (cf. Table 3.11). Thus, for ThF_3 and PaF_3 the assumption of a near-integral 5f occupation is too crude [20], and all m.a.e. and m.r.e. were calculated neglecting the results for these two systems.

Actinide Trifluoride Structure Using the trivalent 5f-in-core LPPs [20] in connection with CPPs, the LPP HF An–F bond lengths and F–An–F bond angles decrease on average by 0.015 Å (0.7%) and 1.5° (1.2%), respectively. Since the SPP state-averaged MCSCF bond lengths and angles are overestimated by about 0.020 Å (0.9%) and 3.7° (3.2%), respectively, using pure LPPs, the application of CPPs reduces the m.a.e. (m.r.e.) to 0.010 Å (0.5%) and 2.1° (1.9%). Thus, the use of CPPs clearly improves the results of the LPPs. In comparison to the Cm–F bond length (R_e=2.095 Å) from an AE DHF calculation [73] the LPP+CPP HF result is also considerably better than the pure LPP HF result, i.e. the LPP and LPP+CPP bond lengths are by 0.019 and 0.001 Å too long, respectively.

If correlation effects are taken into account via CCSD(T), the LPP bond lengths and angles are reduced by about 0.017 Å (0.8%) and 4.5° (3.8%) with respect to the HF values, respectively. The decrease in CCSD(T) bond lengths and angles due to the application of CPPs is almost by 50% smaller than that for the HF calculations, i.e. it amounts to 0.009 Å (0.4%) and 0.8° (0.7%), respectively. The reason for this smaller decrease is most likely that the CCSD(T) bond lengths and angles are already reduced by valence correlation effects and that consequently the further reduction is complicated due to the increased F–F repulsion.

The LPP/LPP+CPP CCSD(T) bond length and angle of UF_3 show reasonable agreement with those from a SPP DFT/PBE0 calculation (R_e=2.069 Å; $\angle$=105°) [76], i.e. the deviations amount to 0.080/0.074 Å (3.9/3.6%) and 7°/6° (6.7/5.7%), respectively. The same is true for the DFT and MP2 results for UF_3 published by Joubert and Maldivi [79], who included scalar-relativistic corrections either by a frozen-core approximation with a quasirelativistic treatment of the valence electron shells or by an energy-adjusted quasirelativistic pseudopotential, where the 5f, 6s, 6p, 6d, and 7s elec-

Table 3.10: An–F bond lengths R_e (in Å), bond angles $\angle$F–An–F (in deg), and ionic binding energies ΔE_{ion} (in eV) for AnF$_3$ (An=Ac–Lr) from HF calculations using LPPs with and without CPPs as well as from SPP state-averaged MCSCF calculations.

An	R_e			$\angle$F–An–F			ΔE_{ion}		
	LPP	CPP[a]	SPP	LPP	CPP[a]	SPP	LPP	CPP[a]	SPP
Ac	2.207	2.200	2.213	115.7	115.4	115.9	42.99	43.18	42.92
Th	2.193	2.184	2.125	115.1	114.7	101.8	43.36	43.59	46.53
Pa	2.179	2.170	2.109	115.1	114.6	113.6	43.71	43.99	45.33
U	2.166	2.154	2.124	115.4	114.7	108.5	44.06	44.40	44.88
Np	2.152	2.139	2.118	115.8	114.9	110.6	44.41	44.81	45.04
Pu	2.139	2.125	2.109	116.2	115.1	112.4	44.75	45.20	45.56
Am	2.126	2.110	2.100	116.6	115.3	112.9	45.08	45.58	45.82
Cm	2.114	2.096	2.097	117.2	115.6	114.5	45.41	45.97	45.71
Bk	2.102	2.084	2.085	117.6	115.7	114.7	45.70	46.31	46.24
Cf	2.090	2.071	2.078	118.1	116.0	114.7	46.03	46.67	46.62
Es	2.080	2.060	2.059	118.5	116.2	114.8	46.31	46.98	46.82
Fm	2.069	2.050	2.045	118.9	116.4	113.9	46.58	47.27	47.10
Md	2.058	2.039	2.043	119.5	117.0	116.0	46.90	47.58	47.46
No	2.047	2.031	2.039	119.7	117.6	116.0	47.18	47.83	47.61
Lr	2.037	2.023	2.034	120.0	118.4	117.1	47.47	48.06	47.57

[a] LPP calculations using CPPs.

trons are treated explicitly. Depending on the method the obtained bond lengths and angles are in-between 2.051–2.122 Å and 104.7–118.4°, respectively. Therefore the LPP and LPP+CPP results deviate in a range of 0.027–0.098 and 0.021–0.092 Å as well as from −6.9 to +6.8 and from −7.2 to +6.5°, respectively. Compared to the Am–F bond length (R_e=2.078 Å) from an AE DFT/BP calculation including the 5f, 6s, 6p, 6d, 7s, and 7p orbitals in the valence space and using ZORA [80], the LPP and LPP+CPP CCSD(T) bond lengths deviate just by 0.031 (1.5%) and 0.022 Å (1.1%), respectively. Furthermore, the differences between the LPP/LPP+CPP CCSD(T) Am–F bond lengths and those from CASPT2 calculations using either a scalar Douglas–Kroll–Hess (DKH) Hamiltonian or the SPP [80] are 0.046/0.037 and 0.027/0.018 Å, respectively (R_e: DKH 2.063; SPP 2.082 Å). Thus, the LPP CCSD(T) results show good agreement with the available data from the literature and the CPPs improve the pure LPP results.

3.1 5F-IN-CORE PSEUDOPOTENTIALS FOR ACTINIDES

Ionic Binding Energy In the case of the LPP HF ionic binding energies the use of CPPs causes a mean increase by 0.50 eV (1.1%). Therefore the deviations from the SPP HF results become more than 50% smaller, if CPPs are applied, i.e. the m.a.e. (m.r.e.) amount to 0.51 (1.1%) and 0.24 eV (0.5%) for LPP and LPP+CPP HF calculations, respectively.

Table 3.11: An–F bond lengths R_e (in Å), bond angles ∠F–An–F (in deg), and ionic binding energies ΔE_{ion} (in eV) for AnF$_3$ (An=Ac–Lr) from CCSD(T) calculations using LPPs with and without CPPs. Additionally, LPP HF and SPP state-averaged MCSCF f orbital populations from a Mulliken population analysis are given.

	R_e		∠F–An–F		ΔE_{ion}		f occ.	
An	LPP	CPP[a]	LPP	CPP[a]	LPP	CPP[a]	LPP[b]	SPP
Ac	2.189	2.185	112.5	112.5	43.82	43.91	0.13	0.23
Th	2.176	2.172	111.7	111.6	44.18	44.30	0.10	0.60
Pa	2.163	2.158	111.4	111.2	44.54	44.68	0.09	1.52
U	2.149	2.143	111.5	111.2	44.89	45.07	0.08	3.02
Np	2.135	2.128	111.7	111.3	45.24	45.45	0.07	4.05
Pu	2.122	2.114	112.0	111.4	45.57	45.82	0.07	5.07
Am	2.109	2.100	112.2	111.5	45.90	46.19	0.06	6.07
Cm	2.097	2.086	112.6	111.7	46.22	46.56	0.06	7.07
Bk	2.085	2.074	112.9	111.7	46.51	46.89	0.06	8.07
Cf	2.073	2.061	113.2	112.0	46.83	47.25	0.05	9.07
Es	2.062	2.050	113.5	112.1	47.11	47.56	0.05	10.07
Fm	2.052	2.040	113.7	112.3	47.39	47.86	0.05	11.06
Md	2.041	2.029	114.2	112.7	47.70	48.18	0.05	12.06
No	2.031	2.020	114.5	113.2	47.99	48.46	0.05	13.05
Lr[c]	2.020	2.012	114.9	113.6	48.28	48.72	0.05	14.04

[a] LPP calculations using CPPs.
[b] 0–14 electrons in the 5f shell are attributed to the LPP core for Ac–Lr, respectively.
[c] SPP CCSD(T): R_e=2.002 Å, ∠F–Lr–F=111.5°, ΔE_{ion}=48.83 eV.

At the CCSD(T) level the LPP HF ionic binding energies are increased by about 0.82 eV (1.8%). Compared to the HF results the application of CPPs yields a smaller mean increase of the ionic binding energies by 0.32 eV (0.7%). Because of this increase, a clear improvement due to CPPs is found in comparison to the SPP CCSD(T) optimization of LrF$_3$ (R_e=2.002 Å; ∠=111.5°; ΔE_{ion}=48.83 eV). The deviations from the SPP CCSD(T) ionic binding energy amount to 0.55 (1.1%) and 0.11 eV (0.2%) for LPP and LPP+CPP calculations, respectively. Furthermore, the LPP+CPP bond lengths and angles are also in better agreement with the SPP data than the pure LPP results, i.e. the differences between SPP and LPP/LPP+CPP results are 0.018/0.010 Å and 3.4/2.1° for bond lengths and angles, respectively. Hence, the use of CPPs also leads to an improvement of the ionic binding energies.

3.1.5.7 Comparison of the Actinide Fluorides

Figure 3.5 shows the deviations between the SPP state-averaged MCSCF and integral or assumed LPP 5f occupations for AnF_n (n=2–6) [20,29,30]. The deviations between these 5f occupations increase with increasing oxidation state, i.e. the m.a.e. amount to 0.01, 0.12, 0.22, 0.59, and 1.35 electrons for di-, tri-, tetra-, penta-, and hexavalent actinide atoms, respectively. The reason for this is that the higher the assumed oxidation state, the less probable it is, since ionization energies increase with increasing positive charge. As one can see for ThF_3 (Z=90) and PaF_3 (Z=91) as well as for PuF_2 (Z=94) the integral 5f occupations are larger than the SPP 5f occupations. This is due to the fact that for these actinides the tri- and divalent oxidation states are not preferred (Th) or even not stable (Pa, Pu) in aqueous solution (cf. Table 1.1) [1]. Thus, the tri- and divalent subconfigurations $5f^n$ and $5f^{n+1}$ mix strongly with the corresponding energetically low-lying tetra- and trivalent subconfigurations $5f^{n-1}6d^1$ and $5f^n6d^1$, respectively, yielding smaller SPP 5f occupations than assumed for the LPP core [20]. For all other cases especially the penta- and hexavalent oxidation states the integral 5f occupations are too small.

Figure 3.6 presents the differences between SPP state-averaged MCSCF and LPP HF 5f occupations for AnF_n (n=2–6) [20, 29, 30]. One can see that the f-part of the LPP, which allows for some additional 5f occupation, clearly reduces the deviations for all AnF_n except for those cases, where the integral 5f occupations are larger than those of the SPPs, i.e. for ThF_3, PaF_3, and PuF_2. The m.a.e. are 0.01, 0.02, 0.07, 0.19, and 0.66 electrons for di-, tri-, tetra-, penta-, and hexavalent actinide atoms, respectively (divalent m.a.e. without Pu; trivalent m.a.e. without Th, Pa). The 5f occupations for AnF_2, AnF_3, and AnF_4 differ at most by ca. 0.1 electrons except for ThF_3, PaF_3, and PuF_2. For AnF_5 the differences in the 5f occupations are slightly larger and differ by at most ca. 0.3 electrons. However, this difference is still acceptable, because the pentavalent PPs yield analogous to the di-, tri-, and tetravalent LPPs reasonable results for

3.1 5F-IN-CORE PSEUDOPOTENTIALS FOR ACTINIDES

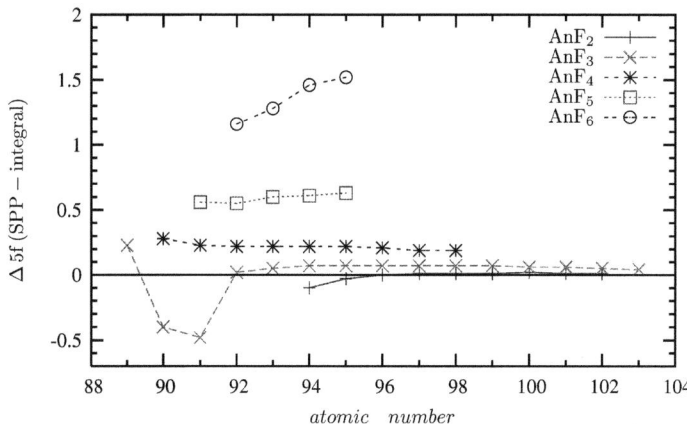

Figure 3.5: Differences between SPP state-averaged MCSCF and integral LPP 5f occupations for AnF_2 (An=Pu–No) [29], AnF_3 (An=Ac–Lr) [20], AnF_4 (An=Th–Cf) [29], AnF_5 (An=Pa–Am) [30], and AnF_6 (An=U–Am) [30].

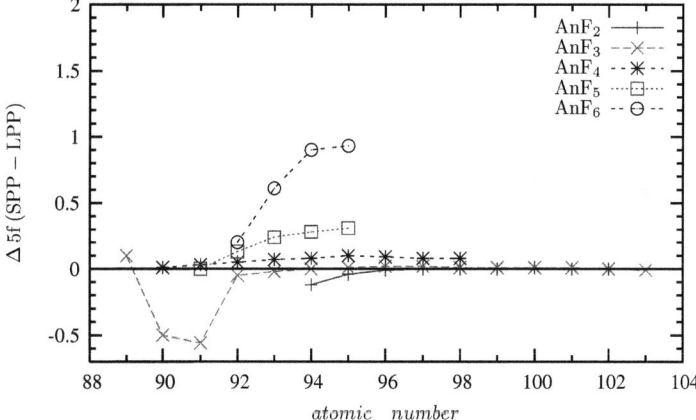

Figure 3.6: Differences between SPP state-averaged MCSCF and LPP HF 5f occupations for AnF_2 (An=Pu–No) [29], AnF_3 (An=Ac–Lr) [20], AnF_4 (An=Th–Cf) [29], AnF_5 (An=Pa–Am) [30], and AnF_6 (An=U–Am) [30].

bond lengths and energies[5], i.e. the m.a.e. (m.r.e.) are at most 0.023 Å (1.1%) and 0.11 eV (3.0%) for bond lengths and energies, respectively (divalent m.a.e. without Pu; trivalent m.a.e. without Th, Pa). For AnF_6, however, the deviations of the 5f occupations are significantly increased up to ca. 0.9 electrons, which explains the failure of the hexavalent PPs in the case of the bond energies. For UF_6 one can see that the deviation of the 5f occupation is comparable to those for AnF_5, wherefore the hexavalent PP for uranium yields reasonable results even for the bond energy ($\Delta 5f$=0.20 electrons, ΔE_{bond}=0.29 eV (5.2%)). Finally, one can conclude that differences between SPP and LPP 5f occupations higher than 0.5 electrons become too large.

3.1.6 Range of Applications

The 5f-in-core LPPs simplify electronic structure calculations on actinide compounds significantly. However, the assumption of a fixed near-integral 5f occupancy also bears the danger of misuse of the approach, e.g., for cases where another 5f occupancy than modeled by the PP is actually present, cases where states with different 5f occupancies mix, or systems where the 5f orbitals strongly contribute directly to chemical bonding in a MO-LCAO (molecular orbitals by linear combination of atomic orbitals) sense. Thus, users of the 5f-in-core LPPs have to verify the underlying assumption by (single-point) test calculations using, e.g. 5f-in-valence SPPs [15, 21] or AE methods at the HF level. It is clear that questions related to individual electronic states cannot be addressed with the present approach, which rather provides answers for an average over a multitude of states characterized by the same 5f occupancy and the same valence substate, i.e. a superconfiguration in the sense of the concept of Field advocated for lanthanides more than two decades ago [81].

The range of possible applications of the actinide 5f-in-core PPs is certainly somewhat smaller than for lanthanide 4f-in-core PPs [17]. Nevertheless, a quite significant part of actinide chemistry remains open for applications of this approach. Possible applications of the divalent 5f-in-core PPs are, e.g., the study of metal clusters of heavier actinides, similar to previous related work on ytterbium clusters [82]. In the case of the tetravalent actinides the bis-cyclooctatetraene complexes have been successfully investigated with 5f-in-core PPs, which were found to be able to model quite well the contributions of f and d orbitals to metal–ring bonding [83] (cf. Sect. 3.1.7). In addi-

---

[5]Bond energies for AnF_3 (An=Ac–Lr) from LPP HF and SPP state-averaged MCSCF calculations [19, 20] are given in Table C.7.

tion a couple of applications have been published for trivalent 5f-in-core PPs, i.e. DFT studies on actinide(III) motexafin complexes (An–Motex^{2+}, An=Ac, Cm, Lr) [23] and on the hydration behavior of trivalent actinide ions [24] demonstrate that the 5f-in-core approach performs encouragingly well. Furthermore, the cohesive energy of crystalline uranium nitride and its electron charge distribution has been investigated using the trivalent 5f-in-core PP [25]. Therefore despite the widespread common knowledge that the actinide 5f shell is chemically active and cannot be attributed to the core, ample quantitative evidence is found that such an approximation can be made without too much loss of accuracy for many cases. However, for the higher, namely penta- and hexavalent, oxidation states the successful applications will noticeably decrease compared to those of the lower oxidation states, because the higher oxidation states are only formally realized in molecules. For example in the case of penta- AnO_2^+ and hexavalent AnO_2^{2+} actinyl ions the 5f-in-core PPs failed except for UO_2^{2+} [30] (cf. Sect. 3.1.8). The investigation of uranyl(VI) complexes with aromatic acids in aqueous solution using the hexavalent LPP showed reasonable results [84] (cf. Sect. 3.1.9). However, the hexavalent uranium PP is no 5f-in-core but only a large-core PP, because its assumed 5f occupation is zero ($5f^0$).

3.1.7 Application to Actinocenes

The accurate treatment of metal sandwich compounds is still a considerable challenge for ab initio quantum chemistry. It was shown to be rather difficult to account accurately for relativistic and electron correlation effects in large organometallic systems like ferrocene [85]. Due to the larger number of atoms and electrons, the greater importance of relativistic effects due to the heavier central atom, as well as the need of higher angular momentum basis functions for the central atom, actinocenes are even more difficult to treat accurately at an ab initio level than ferrocene.

Uranocene was the first of the f-element sandwich complexes and has been synthesized [86] in 1968 after its theoretical prediction [87] in 1963. Soon after, the syntheses of the analogous actinocenes of Th [88], Pa [89], and Np as well as Pu [90] were reported. Although all actinocenes were studied spectroscopically, molecular structures have only been determined for thorocene and uranocene from three-dimensional X-ray diffraction [91]. The actinide ion, which has a formal +4 oxidation state, was found to be sandwiched by two eight-membered aromatic rings $C_8H_8^{2-}$, which are eclipsed in conformation giving the molecule D_{8h} symmetry (cf. Fig. 3.7).

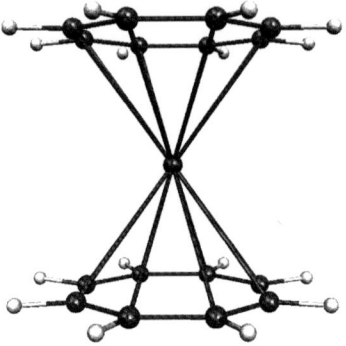

Figure 3.7: Molecular structure of An(C$_8$H$_8$)$_2$ (An=Th–Pu).

One reason why these fascinating complexes are studied extensively by both experimentalists as well as theoreticians is the speculation on 6d and 5f orbital participation in metal–ring bonding. Assuming a completely ionic model, i.e. An^{4+} complexed by two aromatic C$_8$H$_8^{2-}$ ligands, the highest and next-highest occupied ligand orbitals have π character and transform according to the e$_{2u}$ and e$_{2g}$ representations of the D_{8h} point group, respectively. The actinide central ions have low-lying empty 7s (a$_{1g}$), 7p (a$_{2u}$, e$_{1u}$), 6d (a$_{1g}$, e$_{1g}$, e$_{2g}$), and partly occupied 5f (a$_{2u}$, e$_{1u}$, e$_{2u}$, e$_{3u}$) orbitals. Therefore the most important covalent contributions to actinide–ring bonding arise from the actinide 6d (e$_{2g}$) and 5f (e$_{2u}$) orbitals interacting with the ligand π orbitals of the same symmetry [92, 93].

This has experimentally been proven, e.g., by the photoelectron spectrum of uranocene [94], where the e$_{2u}$ ligand band shows a much slower falloff in cross-section as the photo energy is increased than expected for a pure C 2p based ligand band, and responds to the delayed maximum in the metal f band cross-section by a small maximum at 39 eV. The strong resonance in the 90–125 eV region also indicates a large covalent contribution of the f orbitals to the e$_{2u}$ ligand orbitals. Furthermore, the crucial role of 6d orbitals in complex stabilization is implicit in the assignment of the e$_{2u}$ and e$_{2g}$ ligand bands, since the ionization energy of the e$_{2g}$ compared to that of e$_{2u}$ is significantly higher, and thus the e$_{2g}$ orbitals are a larger source of bonding in uranocene.

The extent of the 5f and 6d orbital contributions to actinide–ring bonding has been discussed by several theoreticians. While some studies demonstrated on the basis of

Mulliken orbital populations the primary role of the 6d orbitals [95, 96], others determined a more equal bonding role for the 6d and 5f orbitals [92, 93, 97, 98]. Boerrigter et al. [97] found that the 6d effects are more pronounced in the early actinocenes, with 5f interactions increasing in importance across the series. According to this 5f-in-valence SPP studies on thorocene [96] and uranocene [98] showed the thorocene ground state $5f^0\pi^4$ configuration to be well separated from other configurations and that of uranocene $5f^2\pi^4$ to mix slightly with $5f^3\pi^3$ and $5f^4\pi^2$. Moreover, the seven 5f orbitals in uranocene can be divided into two subgroups, one of which has only one orbital symmetry (e_{2u}) and is considerably delocalized (U character about 90%), while the other one has different orbital symmetries (e_{1u}, e_{3u}, a_{2u}) and is almost completely localized (U character above 99%) [98]. The unpaired f electrons prefer to occupy the localized 5f orbitals, since the 5f–ligand-π interaction raises the energy of the 5f e_{2u} orbitals (SCF orbital energies: -0.396 (e_{2u}), -0.427 (e_{1u}), -0.428 (e_{3u}), -0.433 a.u. (a_{2u})) [95]. The actinocenes are thus interesting candidates to test if such a delicate balance between 6d and 5f orbital participation in chemical bonding can be covered by the 5f-in-core approximation [29], although this approach does not allow configurational mixing between states with different 5f occupancies and models a spherical symmetric 5f shell.

3.1.7.1 Computational Details

The actinocene An(C_8H_8)$_2$ (An=Th–Pu) geometries were completely HF optimized using tetravalent 5f-in-core LPPs [29] with and without CPPs imposing D_{8h} symmetry. For comparison also state-averaged MCSCF calculations using 5f-in-valence SPPs [15] were performed. In the case of thorocene and uranocene LPP calculations using various correlation methods, i.e. MP2 with and without CPPs, CCSD(T), and DFT, were applied, since here experimental structures [91] are available. The 1s orbitals of the 16 C-atoms were frozen in MP2 as well as CCSD(T) calculations. For An energy-optimized (7s6p5d2f1g)/[6s5p4d2f1g] [29] and (14s13p10d8f6g)/[6s6p5d4f3g] [21] valence basis sets were used in LPP and SPP calculations, respectively. For C and H Dunning's cc-pVTZ basis sets [62] (10s5p2d)/[4s3p2d] and (5s2p)/[3s2p] were applied. In the CCSD(T) calculations Dunning's cc-pVDZ basis sets [62] (9s4p1d)/[3s2p1d] and (4s1p)/[2s1p] for C and H had to be used due to limitations in computational resources. As functionals in the DFT optimizations [99] both B3LYP [100–105] and PW91 [104–106] were applied. Here, for C and H TZPP basis sets from the TURBOMOLE [99] library without 1f and 1d function, i.e. (10s6p2d)/[6s3p2d] and (5s2p)/[3s2p], were used, respectively.

The ionic metal–ring binding energy of the actinocenes was defined by $\Delta E = E(An^{4+}) + 2 \times E(C_8H_8^{2-}) - E(An(C_8H_8)_2)$, where the actinide ion was assumed to be in the same $5f^{n-1}$ configuration as in the complex.

Tables 3.12 and 3.13 list LPP HF and correlated results (actinide–ring distances and ionic metal–ring binding energies), respectively. Moreover, Table 3.14 shows the Mulliken population analyses from LPP HF and SPP state-averaged MCSCF calculations.

Table 3.12: Actinide–ring distances R_e (in Å) and ionic metal–ring binding energies ΔE (in eV) for $An(C_8H_8)_2$ (An=Th–Pu) from LPP HF calculations with and without using CPPs in comparison to SPP state-averaged MCSCF, quasirelativistic HFS [97], and experimental [91] data.

	R_e					ΔE		
An	LPP	CPP[a]	SPP	HFS	Exp.	LPP	CPP[a]	SPP
Th	2.080	2.067	2.084	2.08	2.004	79.75	80.17	79.95
Pa	2.066	2.053	2.048	2.02		80.16	80.61	80.98
U	2.047	2.033	2.013	1.98	1.924	80.75	81.26	81.92
Np	2.027	2.012	1.995	1.97		81.37	81.94	82.53
Pu	2.008	1.991	1.973	1.96		81.94	82.59	83.18

[a]LPP calculations using CPPs.

3.1.7.2 Actinide–ring Distance

The actinide–ring distances from LPP HF, LPP+CPP HF, and SPP state-averaged MCSCF calculations decrease almost linearly (correlation coefficients: 0.998, 0.997, and 0.989 for LPP, LPP+CPP, and SPP) with increasing nuclear charge by about 0.1 Å, which is due to the actinide contraction. The LPP results are in good agreement with SPP as well as Hartree–Fock–Slater (HFS) [97] reference data, i.e. the m.a.e. and m.r.e. amount to 0.025/0.04 Å and 1.2/2.2% related to SPP/HFS data, respectively. The application of CPPs shortens the actinide–ring distances by ca. 0.01 Å, wherefore the deviations to SPP and HFS values are reduced to 0.015 (0.8%) and 0.03 Å (1.7%), respectively.

The reason why the SPP results are by 0.011 and 0.026 Å shorter than the former published SPP actinide–ring distances for thorocene (2.095 Å) [96] and uranocene (2.039 Å) [98] is the application of different basis sets, i.e. in the case of Th(C$_8$H$_8$)$_2$ (14s13p10d8f6g)/[6s6p5d4f3g] instead of (12s11p10d8f)/[8s7p6d4f] was applied for Th and in the case of U(C$_8$H$_8$)$_2$ basis sets of VTZ instead of VDZ quality were used. Compared to the experimental data [91] the LPP/LPP+CPP HF distances are obviously too long, i.e. 0.076/0.063 and 0.123/0.109 Å for thorocene and uranocene, respectively. However, after inclusion of electron correlation effects at the MP2/CCSD(T) level about 0.1 Å smaller values are obtained, and thus the deviations from experimental values decrease to −0.038/−0.012 and +0.016/+0.042 Å for thorocene and uranocene, respectively. The use of CPPs at the MP2 level almost does not affect the actinide–ring distances, i.e. their decrease amounts at most to 0.003 Å. Hence, the best Th–ring distance is obtained at the CCSD(T) level and it is in excellent agreement with experiment (1.992 versus 2.004 Å). For uranocene the best agreement is achieved using the MP2 method (1.940 versus 1.924 Å). This is most likely based on an error cancellation, because the experimental U–ring distance is overestimated by all methods and MP2 yields a ca. 0.04 Å too small distance for the simpler case of thorocene. The reason why the deviation for the CCSD(T) value of uranocene is nearly four times higher than that of thorocene (−0.012 versus +0.042 Å) might be that the ground state for uranocene is not purely 5f$^2\pi^4$, but consists of a configurational mixture of 5f$^2\pi^4$ (93.7%), 5f$^3\pi^3$ (3.6%), and 5f$^4\pi^2$ (2.7%) [98]. In contrast to this the 5f$^0\pi^4$ ground state of thorocene is found to be well separated from the excited states [96]. Therefore the tetravalent 5f-in-core PPs, which assume a 5f occupation of zero/two electrons for Th/U, yield more accurate results for thorocene than for uranocene. However, the deviations for uranocene are still acceptable, since the 5f$^2\pi^4$ configuration is clearly dominant.

The LPP DFT calculations using the B3LYP functional yield actinide–ring distances which are by 0.040 and 0.101 Å too long compared to the experimental values for thorocene and uranocene, respectively. Since a DFT study on protactinocene [107] shows that the BLYP functional generates slightly too long Pa–ring distances and that the PW91 functional seems to be the best choice for Pa(C$_8$H$_8$)$_2$, calculations using the PW91 functional were performed, too. Going from B3LYP to PW91 the actinide–ring distances decrease by about 0.08 Å resulting in quite small deviations of −0.038 and +0.026 Å for thorocene and uranocene, respectively. Thus, the good performance of PW91 in the case of the actinocenes is confirmed and the applicability of the 5f-in-core

LPPs in DFT calculations is demonstrated.

Table 3.13: Actinide–ring distances R_e (in Å) and ionic metal–ring binding energies ΔE (in eV) for An(C$_8$H$_8$)$_2$ (An=Th, U) from LPP MP2 calculations with and without using CPPs as well as LPP CCSD(T), LPP DFT/B3LYP, and LPP DFT/PW91 calculations in comparison to experimental data [91].

	R_e		ΔE	
Method	Th	U	Th	U
LPP MP2	1.966	1.940	83.45	84.32
LPP+CPP MP2	1.963	1.938	83.47	84.35
LPP CCSD(T)	1.992	1.966	83.56	84.48
LPP DFT/B3LYP	2.044	2.025	80.85	81.65
LPP DFT/PW91	1.966	1.950	83.99	84.66
Exp.	2.004	1.924		

3.1.7.3 Ionic Metal–ring Binding Energy

The ionic metal–ring binding energies increase almost linearly with the nuclear charge of the central actinide atom (correlation coefficients: 0.998, 0.997, and 0.993 for LPP, LPP+CPP, and SPP) by 2.19, 2.42, and 3.23 eV for LPP HF, LPP+CPP HF, and SPP state-averaged MCSCF calculations, respectively. This is due to the fact that the actinide contraction leads to a decreasing actinide–ring distance, which goes along with an increasing dispersion interaction between the rings.

The agreement between LPP HF and SPP state-averaged MCSCF data is quite good, i.e. the LPP ionic metal–ring binding energies are on average by 0.92 eV (1.1%) too small. Analogous to the actinide–ring distances the application of CPPs improves the results (m.a.e. (m.r.e.): 0.48 eV (0.6%)), since the ionic metal–ring binding energies are increased by about 0.52 eV related to the pure LPP calculations.

The inclusion of electron correlation effects via MP2, CCSD(T), and DFT/PW91 yields an increase of the binding energies by ca. 4 eV (5%), while the energy increase due to the DFT/B3LYP method amounts only to ca. 1 eV (1%). Thus, the DFT/B3LYP energies lie by about 3 eV below the values of the other correlation methods, which is in line with the overestimated actinide–ring distance using DFT/B3LYP. As it is the case for the actinide–ring distance, the use of CPPs at the MP2 level has almost no effect on the binding energy, i.e. the values are augmented by at most 0.03 eV.

Table 3.14: Mulliken 6s/7s, 6p/7p, 6d, and 5f orbital populations and atomic charges (Q) on An in An(C$_8$H$_8$)$_2$ (An=Th–Pu) from LPP HF and SPP state-averaged MCSCF calculations. A $6s^2 6p^6 6d^2 7s^2$ ground state valence sub-configuration is considered for An.

An	s		p		d		f		Q	
	LPP	SPP	LPP	SPP	LPP	SPP	LPP^a	SPP	LPP	SPP
Th	2.24	2.48	6.40	6.35	1.87	1.70	0.37	0.72	1.12	0.75
Pa	2.26	2.49	6.40	6.32	1.94	1.71	0.29	1.68	1.11	0.80
U	2.28	2.50	6.42	6.34	1.97	1.68	0.25	2.66	1.08	0.82
Np	2.29	2.51	6.44	6.37	1.99	1.70	0.22	3.63	1.06	0.79
Pu	2.30	2.52	6.43	6.39	1.99	1.70	0.20	4.60	1.07	0.80

^{a}0–4 electrons in the 5f shell are attributed to the LPP core for Th–Pu, respectively.

Mulliken Orbital Populations From the Mulliken population analyses it can be seen that besides the ionic bonding between the tetravalent metal ion An^{4+} and the two C$_8$H$_8^{2-}$ rings, ligand-to-metal donation and therefore covalent bonding is very important, e.g., the charge of the actinide calculated using LPP/SPP is only about +1.09/+0.79 rather than the formal +4. This charge-transfer occurs to all valence orbitals of the An^{4+} ion, whereby 6d and 5f are preferred to 7s and 7p, e.g., in the case of the SPP state-averaged MCSCF calculation for uranocene the donation amounts to 1.68, 0.66, 0.50, and 0.34 electrons for 6d, 5f, 7s, and 7p, respectively. Since the donation to 6d is almost three times larger than that to 5f, earlier findings [95,96] that for the actinide–ring bonding the 6d orbitals play the primary role are confirmed. However, one should be careful to avoid too detailed interpretations from Mulliken population analyses due to their basis set dependence.

The valence s and p shells get slightly more occupied with increasing nuclear charge, e.g., the 7s/7p occupation increases from 0.24/0.40 for Th(C$_8$H$_8$)$_2$ to 0.30/0.43 electrons for Pu(C$_8$H$_8$)$_2$ in the case of the LPP calculations. The valence 6d occupation, however, stays relatively constant, i.e. an average of 1.95 and 1.70 electrons occupation for 6d of all An(C$_8$H$_8$)$_2$ are observed for LPP and SPP calculations, respectively. For the valence 5f orbitals the occupation decreases by 0.17 and 0.12 electrons from Th(C$_8$H$_8$)$_2$ to Pu(C$_8$H$_8$)$_2$ for LPP and SPP calculations, respectively. The increasing valence s, p and the constant valence d respectively decreasing valence f occupations can be attributed to increasing relativistic effects along the actinide series. The direct relativistic effects are dominating for s and p shells resulting in a contraction and stabilization of 7s/7p, whereas for d and f shells the indirect relativistic effects are larger and lead to an expansion and destabilization of 6d/5f.

The SPP 5f occupations differ on average by 0.66 and at most by 0.72 electrons from the integral LPP $5f^{n-1}$ occupations, which demonstrates that the 5f orbitals participate to some extent in the chemical actinide–ring bonding. However, the 5f-in-core approach still yields quite reasonable results, which is most likely due to the f-part of the PPs, which allows for some 5f occupation in addition to the integral $5f^{n-1}$ occupation modeled by the LPP, i.e. the LPP 5f occupations amount on average to 0.27 electrons. Thus, the mean deviation between SPP and LPP 5f occupations is only 0.39 electrons.

3.1.8 Application to Actinyl Ions

The hexavalent uranyl ion UO_2^{2+} is a linear molecule with very short and strong U–O bonds. The chemical stability of these bonds are well-known and account for the omnipresence of UO_2^{2+} in uranium chemistry [108]. The UO_2^{2+} ion has a closed-shell singlet ground state with 12 valence electrons coming from the O 2p and U 5f, 6d, and 7s atomic orbitals [108]. Analogous to the oxygen dimer the O 2p orbitals form σ_g, σ_u, π_g, and π_u molecular orbitals (MOs). The uranium ion possesses two primary valence shells, 5f and 6d, which can form bonds to these MOs due to their symmetry, i.e. 5f σ_u, 5f π_u, 6d σ_g, and 6d π_g (in $D_{\infty h}$). The resulting four highest occupied MOs in UO_2^{2+}, $3\sigma_g$, $3\sigma_u$, $1\pi_g$, and $2\pi_u$, can be viewed as bonding and thus suggest a notional U–O bond order of three [109]. Since the bonding is strongly dependent on the 5f orbitals, the UO_2^{2+} ion is an interesting but critical candidate to test the hexavalent LPP.

The other penta- AnO_2^+ and hexavalent AnO_2^{2+} actinyl ions are investigated as well, because they are beside UO_2^{2+} vital for underpinning the development of nuclear fuel technologies. However, the calculations showed only reasonable results for UO_2^{2+}, wherefore the results will be discussed separately for UO_2^{2+} and the other actinyl ions, respectively [30].

3.1.8.1 Computational Details

For UO_2^{2+} HF and CCSD(T) calculations were carried out with MOLPRO [52] implying $D_{\infty h}$ symmetry using both the LPP and SPP. In order to be sure that the LPP yields the correct linear structure, LPP HF geometry optimizations [99] using C_1 symmetry and different starting points were performed. All these optimizations resulted in a linear structure, which was identified as a true energy minimum by a numerical vibrational frequency analysis. For O Dunning's aug-cc-pVQZ basis set [62,63] was applied and for An (7s6p5d2f1g)/[6s5p4d2f1g] and (14s13p10d8f6g)/[6s6p5d4f3g] [21] valence basis sets were used for LPP and SPP, respectively. In the CCSD(T) calculations the O 1s orbitals were frozen. The ionic binding energy of UO_2^{2+} was defined by $\Delta E = E(U^{6+}) + 2 \times E(O^{2-}) - E(UO_2^{2+})$. Furthermore, a LPP and SPP DFT/B3LYP [100–105] calculation were performed with TURBOMOLE [99] imposing D_{6h} symmetry. Since in TURBOMOLE exchange–correlation energies are numerically integrated on element specific grids, and since no grid for uranium is implemented, the LPP and SPP calculations were carried out employing the cerium and tungsten m5 grid, respectively, by calculating the corresponding CeO_2^{2+} and WO_2^{2+} molecules and by setting the Ce and W nuclear charge and mass to 92 and 238.03 u, respectively. In the case of the SPP DFT calculations segmented contracted (14s13p10d8f6g)/[10s9p5d4f3g] [110] valence basis sets were used.

For AnO_2^+ (An=U–Am) and AnO_2^{2+} (An=Np–Am) HF calculations were performed in MOLPRO [52] implying C_{2v} symmetry using penta- and hexavalent 5f-in-core PPs, respectively. For O Dunning's aug-cc-pVQZ basis set [62,63] was applied and for An (7s6p5d2f1g)/ [6s5p4d2f1g] valence basis sets were used.

Table 3.15: Bond lengths R_e (in Å), ionic binding energies ΔE (in eV), and f orbital occupations for UO_2^{2+} from LPP HF, CCSD(T), and DFT/B3LYP calculations in comparison to corresponding SPP calculations as well as computational data from the literature.

Method	Ref.	R_e	ΔE	f occ.
LPP HF		1.631	176.05	1.74
SPP HF		1.639	181.21	2.23
AE DHF	[111]	1.650		
LPP CCSD(T)		1.668	177.30	1.69
SPP CCSD(T)		1.689	183.81	2.17
SPP CCSD[a]	[112]	1.697		
quasirel. AE CCSD(T)[a]	[113]	1.683		
AE DHF+CCSD(T)[a]	[114]	1.715		
LPP DFT/B3LYP		1.642	180.44	2.04
SPP DFT/B3LYP		1.692	185.85	2.45
SPP DFT/B3LYP	[113]	1.698		
quasirel. DFT/BPVWN[b]	[108]	1.716		

[a] U 5s, 5p, and 5d as well as O 1s orbitals were frozen.
[b] For U 1s–5d and O 1s orbitals the frozen-core approximation was applied.

3.1.8.2 Uranyl(VI) ion

Table 3.15 shows bond lengths, ionic binding energies, and 5f orbital occupations for UO_2^{2+} from LPP HF, CCSD(T), and DFT/B3LYP calculations in comparison to corresponding SPP calculations and computational data from the literature [108, 111–114]. As one can see the LPP and SPP bond lengths are in good agreement, i.e. the LPP underestimates the SPP bond lengths by 0.008 (0.5%), 0.021 (1.3%), and 0.050 Å (3.0%) at the HF, CCSD(T), and DFT/B3LYP level, respectively. In comparison to the computational data from the literature, the LPP bond lengths differ in a range of 0.015–0.074 Å corresponding to 0.9–4.3%. However, these deviations are not necessarily due to the different core definitions, but may also arise from the use of different basis sets, relativistic approaches, or density functionals. For example, in the case of the largest deviation, which occurs between the LPP DFT/B3LYP and the quasirelativistic DFT/BPVWN [108] calculation, a 1s–5d instead of a 1s–5f core, pVTZ instead of aug-cc-pVQZ basis sets, and the BPVWN instead of the B3LYP density functional were used for the quasirelativistic DFT calculation.

For the ionic binding energies the LPP underestimate the SPP data by 5.16 (2.8%), 6.51 (3.5%), and 5.41 eV (2.9%) for HF, CCSD(T), and DFT/B3LYP calculations, respectively. These small deviations as well as those for the bond lengths can be understood by the comparison of the LPP and SPP 5f orbital occupations, which show deviations below 0.50 electrons for all calculations (0.49, 0.48, and 0.41 electrons for HF, CCSD(T), and DFT/B3LYP). Thus, analogous to the test calculation on UF_6 in the case of the uranyl(VI) ion the hexavalent LPP for uranium ($5f^0$) yields reasonable results.

3.1.8.3 Actinyl(V) and Actinyl(VI) Ions

Table 3.16 lists bond lengths and angles for AnO_2^+ (An=U–Am) and AnO_2^{2+} (An=Np–Am) from LPP HF calculations in comparison to scalar-relativistic AE DFT/PBE [115] and SPP state-averaged MCSCF [116] calculations from the literature, respectively. In contrast to the AE and SPP reference data the LPP HF calculations did not yield linear structures, but bent structures with O–An–O bond angles between 102.2 and 109.3°. Furthermore, the LPP HF bond lengths are by about 0.029 (1.6%) and 0.112 Å (7.0%) longer than those of the AE DFT/PBE [115] and SPP MCSCF [116] calculations for AnO_2^+ and AnO_2^{2+}, respectively. Thus, in the case of these systems the a priori assumption of penta- and hexavalent actinides and a corresponding near-integral 5f occupancy

Table 3.16: Bond lengths R_e (in Å) and angles ∠ O–An–O (in deg) for AnO_2^+ (An=U–Am) and AnO_2^{2+} (An=Np–Am) from LPP HF calculations in comparison to scalar-relativistic AE DFT/PBE [115] and SPP state-averaged MCSCF calculations [116] from the literature, respectively.

	AnO_2^+			AnO_2^{2+}		
	LPP		AE^a	LPP		SPP^a
An	R_e	∠	R_e	R_e	∠	R_e
U	1.776	109.3	1.770			
Np	1.784	106.0	1.750	1.710	108.0	1.628
Pu	1.781	104.7	1.734	1.727	103.7	1.609
Am	1.776	104.0		1.731	102.2	1.594

$^a D_{\infty h}$ symmetries, i.e. bond angles correspond to 180°.

fails. For the hexavalent actinyl ions this failure is in line with the results for AnF_6, where the 5f-in-core approach was too crude as well. However, the LPP test calculations for AnF_5 showed good agreement with SPP reference data, wherefore the bad description of the pentavalent actinyl ions was not expected.

In order to understand the discrepancies between the AE and 5f-in-core PP results for AnO_2^+, a SPP state-averaged MCSCF geometry optimization [52] for UO_2^+ distributing one electron in the seven U 5f orbitals and optimizing the mean energy of the corresponding seven states was performed (cf. Table 3.17; basis sets: U (14s13p10d8f6g)/ [6s6p5d4f3g] [21], O aug-cc-pVQZ [62, 63]; symmetry: C_{2v}). Analogous to the LPP HF result the SPP MCSCF structure is bent with an O–U–O bond angle of 152.1°. If the UO_2^+ structure is optimized for the individual states, four linear and three nonlinear structures are obtained (cf. Table 3.17). From a Mulliken population analysis of the singly occupied molecular orbitals (SOMOs), which are dominantly of U 5f character, it can be seen that the seven U 5f orbitals can be divided into two subgroups. Four of them are pure f orbitals (100% f character) and three have dominant f contributions, but mix with U d and O p orbitals (f character about 75–88%). The pure f orbitals are non-bonding f_δ and f_ϕ orbitals and when singly occupied the corresponding optimizations yield linear structures. The other f orbitals correspond for a linear structure to f_π and f_σ orbitals, which can mix with U d and O p orbitals due to their symmetry. If these orbitals are singly occupied the corresponding optimizations yield bent structures, which have 2–3 eV higher energies than the linear ones with f_δ or f_ϕ singly occupied.

The reason why in C_{2v} the optimization for the average of the seven states yields a bent

Table 3.17: Bond lengths R_e (in Å), bond angles $\angle$O–U–O (in deg), and occupations of the SOMOs, which are dominantly of U 5f character, for UO_2^+ from SPP state-averaged MCSCF geometry optimizations for the energies of the individual states arising from a $5f^1$ occupation as well as for the mean energy of these states $\overline{E}$. Furthermore, the relative energies ΔE with respect to the lowest state $^2\Delta_u$ (in eV) and the energy difference ΔE_{linear} (in eV) to a $D_{\infty h}$ optimized structure is given.

Optimized for	R_e	$\angle$	SOMO occ.	ΔE	ΔE_{linear}^a
$E(^2\Delta_u)$	1.700	180.0	1.00	0.00	
$E(^2\Phi_u)$	1.711	180.0	1.00	0.06	
$E(^2A_1)$	1.725	155.8	0.75	2.26	0.07
$E(^2B_1)$	1.736	117.8	0.88	2.03	0.30
$E(^2B_2)$	1.750	100.1	0.86	3.09	3.63
$\overline{E}$	1.729	152.1		1.66	0.14

$^a \Delta E_{linear} = E(D_{\infty h}) - E(C_{2v})$

structure, although there are more linear than non-linear geometries, is most likely that for the $^2B_2$ state the energy difference between the linear and the bent equilibrium structure is so high (3.63 eV) that the optimization of the mean energy $\overline{E}$ is dominated by this contribution and dragged to a bent structure. Since the pentavalent 5f-in-core PP for uranium describes the average of all $5f^1$ states, a bent UO_2^+ structure is obtained in qualitative agreement with the 5f-in-valence SPP result. However, it is obvious from the SOMO population (cf. Table 3.17) that the assumption of a 5f occupancy of at least one electron is not fulfilled. Thus, UO_2^+ cannot be treated within the 5f-in-core approximation using the pentavalent PP for uranium.

However, if the hexavalent PP for uranium is applied to calculate UO_2^+ by explicitly distributing one electron in the seven f orbitals and optimizing the mean energy of the seven states, a similar structure as in the SPP MCSCF calculation with a bond angle of 160.5° (R_e=1.667 Å) is obtained. Here, the uranium basis set has been slightly increased by using four instead of two f exponents, which have been optimized in state-averaged MCSCF calculations [52] for the $5f^36d^17s^2$ valence subconfiguration of the hexavalent uranium LPP (exponents: 5.719, 2.062, 0.797, 0.266).

If HF instead of state-averaged MCSCF calculations are performed [52] by assuming the single electron to be, e.g., in a f_δ orbital, both SPP and LPP yield linear structures with a bond length difference of 0.058 Å (R_e: SPP 1.700, LPP 1.642 Å). Therefore reasonable results are obtained for the UO_2^+ molecule with the LPP presented here,

3.1 5F-IN-CORE PSEUDOPOTENTIALS FOR ACTINIDES

but the hexa- instead of the pentavalent PP for uranium has to be used. It should be noted, however, that the hexavalent PP for uranium is of large-core, but not really of 5f-in-core type ($5f^0$ occupation).

The other actinyl ions AnO_2^+ are similar to UO_2^+ and the failure of the pentavalent LPPs should therefore also be due to the fact that states, where a f_π or f_σ orbital is occupied, yield non-linear structures. In summary the application of the 5f-in-core approach to these systems cannot be recommended.

3.1.9 Application to Uranyl(VI) Complexes

Natural organic material and microbes can influence the migration behavior of actinides, particularly of uranium, in the environment. For the assessment of risks connected, e.g., with long-term nuclear waste disposal, knowledge of the binding modes of uranium under the various environmental conditions is important. Therefore it is necessary to investigate the complex formation of uranium with selected bioligands, particularly with siderophores of the pyoverdin type, which have a high potential to bind actinides [117–124].

The preferred actinide binding functionalities of pyoverdins are the hydroxamic acid group at the peptide chain and the catechol group of the chromophore [119, 125]. In UV–vis [126] and time-resolved laser-induced fluorescence spectroscopy [127] studies of the complexation behavior of this hydroxamic acid group, salicylhydroxamic acid $HOC_6H_4CONHOH$ (Hsha) and benzohydroxamic acid $C_6H_5CONHOH$ (Hbha) were used as simple model ligands. As a comparison to the hydroxamic acids, benzoic acid C_6H_5COOH (Hba) was also investigated. It was found that the ligands sha$^-$ and bha$^-$ form 1:1 as well as 1:2 (metal ion:ligand) complexes, while the ba$^-$ ligand always yields the 1:1 complex [126]. The strength of the complex formation decreases from sha$^-$ via bha$^-$ to ba$^-$ and the 1:2 complexes are more stable than the corresponding 1:1 complexes. Furthermore, the absorption and fluorescence properties of the uranium(VI) species formed with sha$^-$, bha$^-$, and ba$^-$ were determined. If the uranyl ion UO_2^{2+} is coordinated by the hydroxamates sha$^-$ or bha$^-$, a blue shift of the absorption maximum is observed, whereas coordination to the carboxylate ba$^-$ results in a red shift of the absorption maximum with respect to the spectrum of the "free", i.e. water-coordinated, UO_2^{2+} [126].

The structural data of actinide hydroxamate and benzoate species are scarce. Although sha$^-$ is mainly discussed as coordinating metal ions via the two hydroxamic oxygen atoms ([O,O]-mode) [128–132], there is at least one other reasonable coordination mode via the phenolic oxygen and the nitrogen atom ([N,O']-mode) (cf. Fig. 3.9). Extended X-ray absorption fine structure (EXAFS) spectroscopy alone is often not sufficient to solve the structure of uranium(VI) species, particularly in the presence of mainly light backscattering atoms (C, N, O) in the near-order surrounding of the absorbing uranium atom. However, it has been shown [133–141] that a combination of EXAFS spectroscopy and quantum mechanical methods is a useful tool not only to investigate structures, but also to estimate complex stabilities in solution.

In this section DFT calculations will be presented, which help to determine precise molecular structures from the EXAFS data for the UO_2^{2+} complexes with the model ligands sha$^-$, bha$^-$, and ba$^-$ and which clarify the coordination mode in the $[UO_2\text{sha}]^+$ complex. Furthermore, calculated relative stabilities and time-dependent DFT (TD-DFT) [142, 143] excitation spectra are compared to previous experimental stability constants [126, 127] and UV–vis spectra [126], respectively. Since all experiments are performed in aqueous solution, solvation effects are carefully addressed within the DFT calculations using both an explicit first hydration sphere for the UO_2^{2+} fragment and a continuum model. After the computational details the investigated UO_2^{2+} complexes with one and two model ligands, i.e. 1:1 [144] and 1:2 [84] complexes, will be discussed separately. In the case of the 1:1 complexes only the 5f-in-valence SPP [15] for uranium was applied, while for the 1:2 complexes both SPP and hexavalent LPP [30] were used.

3.1.9.1 Computational Details

The uranium(VI) complexes $[UO_2L]^+$ and $[UO_2L_2]^6$ (L=sha, bha, ba) were completely geometry optimized (symmetry: C_1) at the DFT/B3LYP level [100–105] using the TURBOMOLE program system [99]. For $[UO_2\text{sha}]^+$ both the [O,O]- and [N,O']-mode were considered, while for $[UO_2\text{sha}_2]$ only the [O,O]-mode was calculated, because this mode is found to be clearly favored for the 1:1 complex. For $[UO_2L_2]$ (L=sha, bha) there are two possible structures, i.e. the nitrogen atoms of the ligands can be located on the same or on opposite sides. Since the deviations for SPP bond lengths, bond angles, and total energies are at most 0.015 Å, 4.6°, and 0.0002 a.u. (cf. Table C.8), respectively, only the complexes with the nitrogen atoms on opposite sides will be discussed.

---

[6]The 1:2 complexes were calculated by D. Weißmann.

3.1 5F-IN-CORE PSEUDOPOTENTIALS FOR ACTINIDES   81

In order to ensure that the obtained structures are true energy minima on the potential energy surface (PES), numerical vibrational frequency analyses were performed. For uranium the 5f-in-valence SPP [15] was used in combination with the (14s13p10d8f)/ [10s9p5d4f] segmented contracted valence basis set [110]. The accuracy of this scalar-relativistic valence-only model for structures and binding energies, if used in combination with hybrid density functionals, was demonstrated, e.g. by Batista et al. [145]. In the case of the 1:2 complexes also the hexavalent LPP [30] was used in combination with the (7s6p5d2f1g)/[6s5p4d2f1g] basis set. All other atoms were treated at the AE level using TZP basis sets from the TURBOMOLE library [99]. Except for the UO_2^{2+} oxygen atoms, all oxygen and nitrogen basis sets were augmented by one diffuse s-, p-, and d-function taken from the aug-cc-pVDZ basis sets [63] (exponents: O: s 0.0790, p 0.0686, d 0.3320; N: s 0.0612, p 0.0561, d 0.2300).

Since in TURBOMOLE [99] exchange–correlation energies are numerically integrated on element specific grids, and since no grid for uranium is implemented, all SPP and LPP calculations were carried out employing the tungsten m5 grid by calculating the corresponding WO_2^{2+} molecules and by setting the W nuclear charge and mass to 92 and 238.03 u, respectively. Total energies were converged to 10^{-9} a.u. Beside the gas phase optimizations, calculations including solvation effects[7] were performed as well (cf. below).

The electronic excitations were treated within the adiabatic approximation of TD-DFT [142, 143] considering the 100 energetically lowest singlet excitations. For every calculated spectral line, one Gaussian function $a_0 \exp[-b(\lambda - \lambda_0)^2]$ was used to represent the contribution to the spectrum. Here a_0 is the oscillator strength, λ_0 is the wavelength, and b=0.005 is a broadening parameter. The continuous spectrum in a given interval was obtained pointwise, as the sum over the 100 Gaussian functions, for each wavelength λ.

Modeling Solvation Within a stationary quantum chemistry framework, one can address solvation effects on molecular and electronic structures by applying continuum models [146] or by modeling parts of or complete discrete solvation shells around the solute system of interest (levels 1 and 2, respectively, cf. Fig. 3.8). Moreover, one can combine both in a hybrid-type level 3 approach, modeling solute–solvent clusters for gas phase conditions and correcting for long-range interactions by single-point energy

---

[7]For the 1:1 complexes the calculations including solvation effects were performed by J. Wiebke.

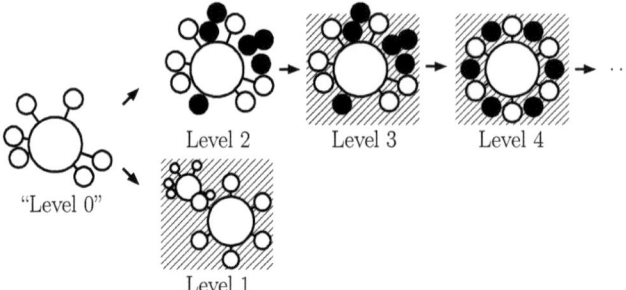

Figure 3.8: Different levels of addressing solvation effects. Note that there is no well-defined hierarchy among the levels 1 and 2.

calculations applying a continuum model. A level 4 approach would additionally involve molecular structure relaxation within the continuum model potential.

Solvation effects on the $[UO_2L]^+$, $[UO_2L_2]$, and UO_2^{2+} systems have been modeled employing a hybrid-type level 3 approach. It has been shown in recent investigations of UO_2^{2+} hydration [147–150] that this is a reliable strategy with inherent, but known shortcomings, which can be identified within a well-defined hierarchy of increasingly better approximations to aqueous solution conditions (cf. Fig. 3.8). Moreover, within a full level 4 approach, one encounters poor molecular structure convergence [149] or convergence to PES saddle points only [147], which has been recently discussed to be an artifact due to the discretization of the solute cavity employed by the continuum models [151].

As continuum model the Conductor-like Screening Model (COSMO) [152] was used as implemented [153] in the TURBOMOLE program package [99]. The continuum permittivity was set to infinite. Solute cavities were constructed using the default parameter set and the COSMO-RS atomic radii [154] of 1.30, 2.00, 1.83, and 1.72 Å for H, C, N, and O, respectively, and 2.00 Å for U [148]. $[UO_2sha(OH_2)_3]^+$ COSMO screening energies for U atomic radii varying in a range of 1.00–3.00 Å do not deviate from the 2.00 Å radius screening energy by more than 3 kJ mol^{-1}.

Because of computational feasibility, and because the UO_2^{2+} fragment was expected to be the most strongly affected by solvation effects, a complete solvation shell was only modeled for the UO_2^{2+} fragment, i.e. no discrete ligand solvation was considered. For the 1:1 complexes a realistic number x' of OH_2 ligands in the UO_2^{2+} equatorial plane

3.1 5F-IN-CORE PSEUDOPOTENTIALS FOR ACTINIDES

was obtained from considering the rearrangements

$$[UO_2sha(OH_2)_{x-1} \cdot OH_2]^+ \rightarrow [UO_2sha(OH_2)_x]^+ \qquad (3.5)$$

with $[UO_2sha(OH_2)_{x-1} \cdot OH_2]^+$ having $x-1$ OH_2 ligands in the first and one OH_2 ligand in the second UO_2^{2+} solvation shell. Then, x' is the largest x for which the rearrangement (3.5) gives negative reaction energies. Within the level 3 approach, optimal molecular structures without, and total energies within the COSMO potential were calculated for the $[UO_2sha(OH_2)_{x-1} \cdot OH_2]^+$ (x=3, 4) and $[UO_2sha(OH_2)_x]^+$ (x=0, 1, 2, 3) systems. For $[UO_2sha(OH_2)_4]^+$ no PES minimum structure was found, since during attempted structure optimizations the forth OH_2 ligand moved out of the first UO_2^{2+} solvation shell to give $[UO_2sha(OH_2)_3 \cdot OH_2]^+$ in all cases.
For x=3 $[UO_2sha(OH_2)_3]^+$ is favored over $[UO_2sha(OH_2)_2 \cdot OH_2]^+$ by 47 kJ mol^{-1}. Rigorously, from this one can only conclude that x'=3 is the minimum number of OH_2 ligands coordinated to the UO_2^{2+} fragment of $[UO_2sha]^+$. However, x'=3 fits best to experimental EXAFS coordination numbers, which are found to be five as well. Moreover, for x=0, 1, 2, and 3 the $[UO_2sha(OH_2)_x]^+$ mean equatorial U–O$_{eq}$ distances of 2.224, 2.330, 2.392, and 2.460 Å, respectively, approach the experimental EXAFS value of 2.41 Å as x increases. Although the x=3 value is ca. 0.05 Å too large, this is known [147, 155–157] as a systematic level 3 overestimation of metal ion–OH_2 distances due to the neglect of molecular structure relaxation within the COSMO potential, which is the more severe the larger x is. Moreover, this might also be due to the DFT/B3LYP method, which has been shown to overestimate especially equatorial bond distances in PuO_2^{2+} complexes [158]. Therefore $[UO_2sha(OH_2)_3]^+$ is believed to be the most consistent structure model for the uranyl(VI)–sha system. Because of the great similarity of bha$^-$ and ba$^-$ with sha$^-$, and because of the experimental EXAFS data, x'=3 is assumed for the bha and ba systems, too. For $[UO_2ba(OH_2)_3]^+$, however, no PES minimum, but only a first order PES saddle point was found in all cases; following that structure's ω=i 20.8 cm^{-1} eigenmode did not lead to any PES minimum corresponding to a $[UO_2ba(OH_2)_3]^+$ complex. Similar difficulties were stated for PuO_2^{2+} complexes [158] and attributed to the DFT/B3LYP method.
For the 1:2 complexes an analogous procedure was carried out for the $[UO_2sha_2]$ system, i.e. optimal molecular structures without, and total energies within the COSMO potential were calculated for the $[UO_2sha_2(OH_2)_{x-1} \cdot OH_2]$ (x=2) and $[UO_2sha_2(OH_2)_x]$ (x=1, 2) systems. For x=2 $[UO_2sha_2(OH_2) \cdot OH_2]$ is preferred to $[UO_2sha_2(OH_2)_2]$ by

17 kJ mol^{-1}. Thus, in the case of the 1:2 complexes the coordination number is found to be five analogous to the experimental EXAFS results. Therefore these complexes were calculated with one additional OH$_2$ ligand in the UO$_2^{2+}$ equatorial plane.

3.1.9.2 1:1 Complexes

In the following molecular structures, relative stabilities, and excitation spectra of the 1:1 complexes [UO$_2$L]$^+$ (L=sha, bha, ba) from SPP DFT/B3LYP calculations will be compared to experimental data [126, 127, 144]. In the case of the sha system the preferred coordination mode, whether [O,O] or [N,O'], will be determined.

Molecular Structures Figure 3.9 shows the calculated gas phase molecular structures of [UO$_2$L]$^+$ (L=sha, bha, ba). For [UO$_2$sha]$^+$ both the [O,O]- and [N,O']-mode are given. As one can see the [N,O']-mode [UO$_2$sha]$^+$ has a very long U–N bond length and is clearly more distorted than the [O,O]-mode [UO$_2$sha]$^+$, which already suggests that the [N,O']-mode is less stable. In contrast to [UO$_2$bha]$^+$ the ligand of the [O,O]-mode [UO$_2$sha]$^+$ is in-plane. This is most likely due to the phenolic OH group, which can interact with the NH group of the hydroxamic acid functionality and which is not present in the bha system. The ligand of [UO$_2$ba]$^+$ is again in-plane so that this complex has C_S symmetry.

In Table 3.18 calculated gas phase bond lengths and angles for [UO$_2$L]$^+$ (L=sha, bha, ba) and UO$_2^{2+}$ are listed. Generally, the U–O$_{ax}$ distances are found to increase, if bare UO$_2^{2+}$ is complexed. For UO$_2^{2+}$, where the EXAFS sample is pure (100% UO$_2^{2+}$), DFT underestimates the experimental U–O$_{ax}$ distance by ca. 0.07 Å. For the complexes [UO$_2$L]$^+$, however, the DFT U–O$_{ax}$ bond lengths are at most ca. 0.03 Å shorter than those given by EXAFS. These smaller deviations are most likely due to an error cancellation, since the experimental U–O$_{ax}$ bond lengths, which are averaged over all UO$_2^{2+}$ species, appear to short due to the large "free" UO$_2^{2+}$ fraction in the EXAFS samples. Furthermore, a reason for these underestimations is the neglect of solvation effects, because additional water ligands in the UO$_2^{2+}$ equatorial plane increase the U–O$_{ax}$ distances. Thus, the smaller deviations in the case of the complexes can also be explained by the fact that here only three instead of five equatorial water ligands are missing. In the case of the U–O$_{eq}$ distances one should not compare gas phase DFT and EXAFS data, since the gas phase calculations consider UO$_2^{2+}$ only as bi- and not as penta-coordinated. The comparison of the [N,O']-mode [UO$_2$sha]$^+$ with

3.1 5F-IN-CORE PSEUDOPOTENTIALS FOR ACTINIDES

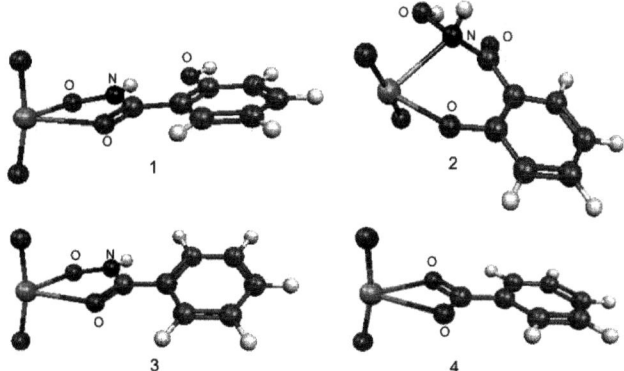

Figure 3.9: $[UO_2L]^+$ (L=sha, bha, ba) gas phase molecular structures. **1, 2, 3,** and **4** correspond to the [O,O]- and [N,O']-mode $[UO_2sha]^+$, to $[UO_2bha]^+$, and to $[UO_2ba]^+$, respectively.

the other complexes confirms the observation from Fig. 3.9 that the U–N distance of 2.395 Å is quite long. The U–O_{eq} bond lengths in the hydroxamate systems [O,O]-mode $[UO_2sha]^+$ and $[UO_2bha]^+$ are nearly the same ($\Delta_{max}R_e$=0.007 Å). The U–O_{eq} distances of the $[UO_2ba]^+$ complex, however, deviate by up to 0.072 Å from those of the hydroxamate systems. A further change due to the complex formation is that the linear O–U–O unit of UO_2^{2+} becomes slightly bent.

Table 3.19 shows the calculated structure parameters for the complexes $[UO_2L(OH_2)_3]^+$ (L=sha, bha) as well as for the solvated uranyl ion $[UO_2(OH_2)_5]^{2+}$. Analogous to the gas phase results, the U–O_{ax} distances in $[UO_2(OH_2)_5]^{2+}$ increase, if two water molecules are substituted by a bidentate hydroxamate ligand. But due to the consideration of the solvent effect, here the EXAFS values are only slightly under- and overestimated for $[UO_2(OH_2)_5]^{2+}$ and $[UO_2L(OH_2)_3]^+$ by ca. −0.02 and at most +0.01 Å, respectively. The comparison of the mean calculated bond lengths $\bar{R}$(U–O_{eq}) for the [O,O]- and [N,O']-mode $[UO_2sha(OH_2)_3]^+$ with the EXAFS data indicates that sha$^-$ is coordinated via the [O,O]-mode, since for this mode $\bar{R}$(U–O_{eq}) exceeds the experimental value only by ca. 0.05 Å as opposed to ca. 0.09 Å for the [N,O']-mode. Furthermore, $\bar{R}$(U–O_{eq}) of the [N,O']-mode $[UO_2sha(OH_2)_3]^+$ is larger than that of $[UO_2(OH_2)_5]^{2+}$, which is due to the very long U–N distance of 2.698 Å. For the [O,O]-mode sha and the bha complex, however, $\bar{R}$(U–O_{eq}) decrease by 0.036 and

Table 3.18: Bond lengths R_e (in Å), angles $\angle$ (in deg), and binding energies E (in kJ mol^{-1}) for the complexes [UO$_2$L]$^+$ (L=sha, bha, ba) and the bare uranyl ion UO$_2^{2+}$ from SPP DFT/B3LYP gas phase calculations compared to EXAFS bond lengths $R_{\text{exp.}}$(U–O$_{\text{ax}}$) [144] and experimental stability constants lg β [126]. For [UO$_2$sha]$^+$ the results for both the [O,O]- and [N,O']-mode are given.

	[UO$_2$sha]$^+$		[UO$_2$bha]$^+$	[UO$_2$ba]$^+$	UO$_2^{2+}$
	[O,O]	[N,O']			
R_e(U–O$_{\text{ax}}$)	1.757	1.760	1.755	1.749	1.698
	1.759	1.760	1.757	1.749	1.698
$R_{\text{exp.}}$(U–O$_{\text{ax}}$)^a	1.77		1.77	1.78	1.77
R_e(U–O$_{\text{Carb.}}$)^b	2.272		2.279	2.247	
R_e(U–ON)	2.175		2.178	2.247^c	
R_e(U–O$_{\text{Ph}}$)		2.114			
R_e(U–N)		2.395			
$\angle$O$_{\text{ax}}$–U–O$_{\text{ax}}$	167.1	170.7	167.4	168.3	180.0
$\angle$O$_{\text{Carb.}}$–U–ON^b	67.5		67.3	58.0^c	
$\angle$N–U–O$_{\text{Ph}}$		71.2			
E	1649	1610	1616	1510	
lg β	17.12		7.96	3.37	

^{a}Composition of the EXAFS samples: sha$^-$: 17% [UO$_2$sha]$^+$, 32% [UO$_2$sha$_2$], 51% UO$_2^{2+}$; bha$^-$: 23% [UO$_2$bha]$^+$, 39% [UO$_2$bha$_2$], 38% UO$_2^{2+}$; ba$^-$: 72% [UO$_2$ba]$^+$, 28% UO$_2^{2+}$; UO$_2^{2+}$: 100% UO$_2^{2+}$.
^{b}O$_{\text{Carb.}}$ is the oxygen atom of the carbonyl group.
^{c}In the case of ba$^-$ the coordinating oxygen atom denoted as ON corresponds to the original hydroxyl group.

0.035 Å compared to [UO$_2$(OH$_2$)$_5$]$^{2+}$, respectively, which is qualitatively in line with the EXAFS results. The fact that EXAFS U–O$_{\text{eq}}$ bond lengths decrease only by 0.01 Å upon coordination is attributed to the large fraction of "free" UO$_2^{2+}$ present in the EXAFS samples. $\bar{R}$(U–O$_{\text{eq}}$) of the [O,O]-mode sha and the bha system differ only by 0.001 Å, which may explain why the EXAFS data show no difference. The overestimation of the experimental U–O$_{\text{eq}}$ distances by 0.05, 0.05, and 0.08 Å for [O,O]-mode [UO$_2$sha(OH$_2$)$_3$]$^+$, [UO$_2$bha(OH$_2$)$_3$]$^+$, and [UO$_2$(OH$_2$)$_5$]$^{2+}$, respectively, is connected with the shortcomings of the applied level 3 solvation model, i.e. the exclusive hydration of the UO$_2^{2+}$ fragment and the neglect of bulk solvation effects. Furthermore, this can be due to the DFT/B3LYP method, which has been shown to overestimate especially equatorial bond distances in PuO$_2^{2+}$ complexes [158]. Here, the deviations for [UO$_2$L(OH$_2$)$_3$]$^+$ are again smaller than that for [UO$_2$(OH$_2$)$_5$]$^{2+}$ due to the large frac-

3.1 5F-IN-CORE PSEUDOPOTENTIALS FOR ACTINIDES

Table 3.19: Bond lengths R_e (in Å) and angles $\angle$ (in deg) for the complexes [UO$_2$L(OH$_2$)$_3$]$^+$ (L=sha, bha) and the solvated uranyl ion [UO$_2$(OH$_2$)$_5$]$^{2+}$ from SPP DFT/B3LYP calculations including solvation effects compared to experimental EXAFS data [144]. For [UO$_2$sha(OH$_2$)$_3$]$^+$ the results for both the [O,O]- and the [N,O']-mode are given.

	[UO$_2$sha(OH$_2$)$_3$]$^+$		[UO$_2$bha(OH$_2$)$_3$]$^+$	[UO$_2$(OH$_2$)$_5$]$^{2+}$
	[O,O]	[N,O']		
R_e(U–O$_{ax}$)	1.774	1.771	1.773	1.749
	1.776	1.775	1.775	1.749
$\bar{R}_e$(U–O$_{ax}$)	1.775	1.773	1.774	1.749
$R_{exp.}$(U–O$_{ax}$)^a		1.77	1.77	1.77
R_e(U–O$_{Carb.}$)^b	2.363		2.372	
R_e(U–ON)	2.284		2.292	
R_e(U–O$_{Ph}$)		2.165		
R_e(U–N)		2.698		
R_e(U–OH$_2$)	2.537	2.516	2.538	2.495
	2.557	2.547	2.550	2.495
	2.560	2.574	2.552	2.496
				2.497
				2.497
$\bar{R}_e$(U–O$_{eq}$)	2.460	2.500	2.461	2.496
$R_{exp.}$(U–O$_{eq}$)^a		2.41	2.41	2.42
$\angle$O$_{ax}$–U–O$_{ax}$	169.1	174.4	169.4	179.8
$\angle$O$_{Carb.}$–U–ON^b	65.8		65.9	
$\angle$N–U–O$_{Ph}$		64.7		

^{a}Composition of the EXAFS samples: sha$^-$: 17% [UO$_2$sha]$^+$, 32% [UO$_2$sha$_2$], 51% UO$_2^{2+}$; bha$^-$: 23% [UO$_2$bha]$^+$, 39% [UO$_2$bha$_2$], 38% UO$_2^{2+}$; ba$^-$: 72% [UO$_2$ba]$^+$, 28% UO$_2^{2+}$; UO$_2^{2+}$: 100% UO$_2^{2+}$.
^{b}O$_{Carb.}$ is the oxygen atom of the carbonyl group.

tion of "free" UO$_2^{2+}$ present in the EXAFS samples.
The O$_{ax}$–U–O$_{ax}$ bond angles of the [UO$_2$L]$^+$ complexes are increased by about 2.6°, if solvation effects are included. Therefore the deviation from the experimentally observed linear O–U–O unit is slightly decreased compared to the gas phase structures of the uranyl complexes.

Relative Stabilities The zero-point corrected uranyl–ligand binding energies (cf. Table 3.18) were obtained by $E = E(\text{UO}_2^{2+}) + E(\text{L}^-) - E([\text{UO}_2\text{L}]^+)$ and the zero-point energies were scaled by 0.972 [159]. Only gas phase binding energies are given, because explicit hydration was exclusively considered for the UO_2^{2+} fragment and not for ligands, resulting in energies that are too small, if calculated via $E = E([UO_2(OH_2)_5]^{2+}) + E(L^-) - E([UO_2L(OH_2)_3]^+) - 2 \times E(OH_2)$.

The comparison between the binding energies of the [O,O]- and the [N,O']-mode $[UO_2sha]^+$ shows that the latter complex is less stable by 39 kJ mol^{-1}. This confirms the conclusion based on the structure data that the sha$^-$ ligand binds via the two hydroxamic acid oxygen atoms. Analogous to the experimental stability constants [126], the calculated binding energies demonstrate that the complex stabilities decrease from [O,O]-mode $[UO_2sha]^+$ via $[UO_2bha]^+$ to $[UO_2ba]^+$.

Electronic Spectra The electronic spectrum of bare UO_2^{2+} has previously been discussed [160]. All transitions in the near-infrared and visible region are dipole-forbidden, but may become allowed via interaction with equatorial ligands. Calculated bare UO_2^{2+} and $[UO_2Cl_4]^{2-}$ excitation energies were found to agree with experimental spectroscopic data for solid-state $Cs_2[UO_2Cl_4]$ within ca. 30 and 11 nm at MRCI (multireference configuration interaction) [161] and CASPT2 [162] levels of theory, respectively. In contrast, however, it has recently been pointed out [163] that one cannot take for granted that excitation energies for the bare UO_2^{2+} calculated at the TD-DFT level are always reliable.

Because of the C_1 symmetry of the systems studied here only trivial (a_1) irreducible representations can be assigned to their electronic states. Therefore the systems (canonical Kohn–Sham) occupied and virtual MOs, which contribute to the electronic excitations, calculated from the TD-DFT response functions will be considered. The MOs have been assigned to the UO_2^{2+}, L^-, and OH_2 subsystems of $[UO_2L(OH_2)_3]^+$ (L=sha, bha) by their MO coefficients and by the Mulliken population analysis.

In Fig. 3.10 the [O,O]-mode $[UO_2sha(OH_2)_3]^+$ excitation spectrum calculated within the COSMO potential is shown as a representative for both the sha and the corresponding bha system. Both systems' excitation spectra show three distinct groups of excitations, which differ in wavelengths λ and relative intensities: large-intensity $UO_2^{2+} \leftarrow OH_2$ charge-transfer (CT) excitations with $\lambda \leq 200$ nm, $L^- \pi^* \leftarrow \pi$ excitations of intermediate intensities in the 200–300 nm region, and $UO_2^{2+} \leftarrow L^-$ CT excitations of low intensities in the 300–700 nm region.

As shown in Fig. 3.11 the experimental uranyl–sha and –bha systems' UV–vis spectra show one broad absorption band in the 350–450 nm region each, with absorption

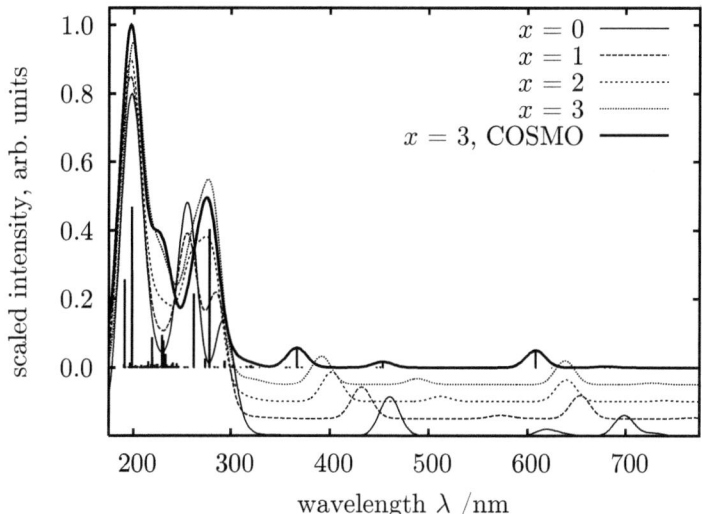

Figure 3.10: Calculated TD-DFT excitation spectra for the [O,O]-mode $[UO_2sha(OH_2)_x]^+$ systems with x=0, 1, 2, 3 and for x=3 within the COSMO potential.

maxima at 402 and 401 nm, respectively [126]. The calculated absorption maxima of the [O,O]-mode [UO$_2$sha(OH$_2$)$_3$]$^+$ and [UO$_2$bha(OH$_2$)$_3$]$^+$ systems are at 367 and 373 nm, i.e. blue-shifted by 35 and 28 nm or 0.29 and 0.23 eV, respectively, from the experimental absorption maxima. In the [O,O]-mode [UO$_2$sha(OH$_2$)$_3$]$^+$ excitation spectrum there is a second low-intensity excitation at 454 nm, i.e. 52 nm or 0.35 eV from the experimental absorption maximum. All these maxima involve CT excitations from L$^-$ π MOs to U 5f atomic-orbital-like MOs. The [O,O]-mode [UO$_2$sha(OH$_2$)$_3$]$^+$ 367 nm and the [UO$_2$bha(OH$_2$)$_3$]$^+$ 373 nm excitations are assigned to the corresponding experimental absorption maxima, because, considering the calculated λ range of ca. 200–700 nm, these match the latter within the expected systematic TD-DFT error, and because experimental UV–vis spectra are dominated by large-intensity intramolecular ligand excitations for λ < 350 nm [126] as calculated. Moreover, when comparing to experimental data one has to consider that the UO$_2^{2+}$–L$^-$ complex formation is far from being quantitative, i.e. that experimental UV–vis spectra might be dominated by the "free", though solvated, UO$_2^{2+}$ ion's absorption bands centered at 414 nm [126].

Compared to the [O,O]-mode the calculated [N,O']-mode [UO$_2$sha(OH$_2$)$_3$]$^+$ excitation spectrum shows no suitable excitations in the 350–450 nm interval. There are L$^-$ π* ← π and UO$_2^{2+}$ ← L$^-$ CT excitations at 322 and 526 nm, being blue- and red-shifted by 80 and 124 nm or 0.73 and 0.77 eV, respectively, from the experimental sha–uranyl system's absorption maximum. Therefore the [N,O']-mode may also be ruled out by comparison of its calculated with the experimental absorption spectrum.

From the calculated excitation spectra of the [O,O]-mode [UO$_2$sha(OH$_2$)$_x$]$^+$ systems, as shown in Fig. 3.10, one can see that accounting for solvation effects is important to bring calculated excitation spectra in line with experimental data. Neglecting solvation effects by setting x=0 apparently gives a correct number and ordering of excitations, which are, however, significantly red-shifted with respect to both x >0 and experimental data. Inclusion of discrete OH$_2$ ligands in the UO$_2^{2+}$ equatorial plane improves the calculated excitation spectra as x increases. If x=3, inclusion of the COSMO potential blue-shifts the UO$_2^{2+}$ ← L$^-$ CT excitation from 391 to 367 nm, which is a large part of the 35 nm or 0.29 eV mismatch of the calculated excitation from the experimental absorption maximum. The fact that improving the solvation model with COSMO, i.e. by going from a level 2 to a level 3 solvation model in Fig. 3.8, causes a more pronounced mismatch of calculated and experimental absorption spectra is believed to point to the solvation model's shortcomings, i.e. the neglect of discrete L$^-$ subsystem solvation

3.1 5F-IN-CORE PSEUDOPOTENTIALS FOR ACTINIDES

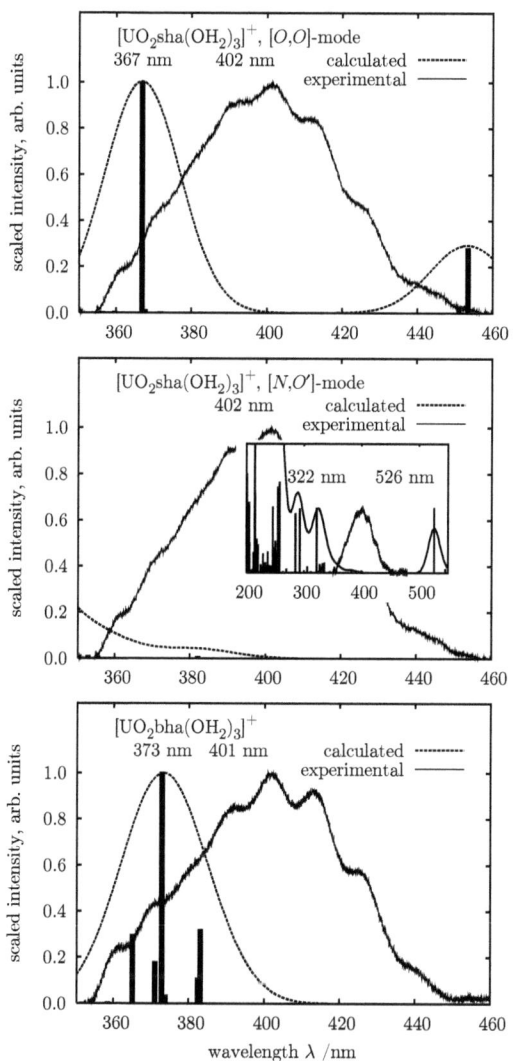

Figure 3.11: Calculated TD-DFT spectra for the [O,O]- and [N,O']-mode $[UO_2sha(OH_2)_3]^+$ and $[UO_2bha(OH_2)_3]^+$ systems in comparison to experimental UV–vis spectra [126] from aqueous solution. Calculated intensities were scaled to have relative intensities of one for the 367, 322, and 373 nm excitations, respectively.

and the level 3 approximation of bulk solvation effects by a simple dielectric model.
In order to address the question of exchange–correlation density functional dependency, the [O,O]-mode $[UO_2sha(OH_2)_3]^+$ excitation spectrum was recalculated using the gradient-corrected PW91 [104–106] and BP86 [102–105, 164] and the hybrid-type PBE0 functionals [104–106, 165, 166]. Gradient-corrected functionals were found to give excitation wavelengths up to 100 nm too large with respect to hybrid-type functionals, whereas the latter give a somewhat qualitatively consistent picture. However, using PBE0 the $[UO_2sha(OH_2)_3]^+$ and $[UO_2bha(OH_2)_3]^+$ absorption maxima were calculated at 396 and 342 nm, whereas the experimental absorption maxima are at 402 and 401 nm, respectively. Thus, there appears to be no systematic functional dependency in quantitative terms.

3.1.9.3 1:2 Complexes

In the following molecular structures, relative stabilities, and excitation spectra of the 1:2 complexes $[UO_2L_2]$ (L=sha, bha, ba) from SPP and LPP DFT/B3LYP calculations [84] will be compared to experimental data [84, 126, 127] except for $[UO_2ba_2]$, because this complex is not observed experimentally.

Molecular Structures In Table 3.20 calculated gas phase bond lengths and angles for $[UO_2L_2]$ (L=sha, bha, ba) and UO_2^{2+} are listed. Analogous to the 1:1 complexes the U–O_{ax} bond lengths are increased, if the bare uranyl is complexed. While the experimental U–O_{ax} bond length for UO_2^{2+} is underestimated by 0.07 Å, the corresponding bond lengths in the complexes are slightly overestimated by at most 0.01 Å and underestimated by ca. 0.05 Å for the SPP and LPP, respectively. These smaller deviations are due to the fact that for the complexes only one equatorial ligand and for the bare uranyl ion five equatorial ligands are missing compared to the penta-coordinated experimental uranyl unit. Again for the U–O_{eq} bond lengths the comparison to the experimental values is not significant, because the complexes are only tetra- and not penta-coordinated. The U–O_{eq} bond lengths of the hydroxamate systems $[UO_2sha_2]$ and $[UO_2bha_2]$ differ only slightly by at most 0.007 Å. Those for the $[UO_2ba_2]$ complex, however, are by up to 0.069 Å longer compared to the U–O_{eq} distances of the hydroxamate systems. In contrast to the 1:1 complexes the O–U–O unit is always found to be linear, which is probably due to the symmetry, i.e. while the 1:1 complexes have C_1 symmetry, the investigated 1:2 complexes have C_i symmetry. The 1:2 complexes with C_1 symmetry,

3.1 5F-IN-CORE PSEUDOPOTENTIALS FOR ACTINIDES

Table 3.20: Bond lengths R_e (in Å), angles $\angle$ (in deg), and binding energies E (in kJ mol^{-1}) for the complexes [UO$_2$L$_2$] (L=sha, bha, ba) and the bare uranyl ion UO$_2^{2+}$ from SPP and LPP DFT/B3LYP gas phase calculations compared to EXAFS bond lengths $R_{\exp.}$(U–O$_{ax}$) [84] and experimental stability constants $\lg\beta$ [126].

	[UO$_2$sha$_2$]		[UO$_2$bha$_2$]		[UO$_2$ba$_2$]		UO$_2^{2+}$
	SPP	LPP	SPP	LPP	SPP	LPP	SPP
R_e(U–O$_{ax}$)	1.782	1.734	1.781	1.732	1.770	1.720	1.698
	1.782	1.734	1.781	1.732	1.770	1.720	1.698
$R_{\exp.}$(U–O$_{ax}$)^a	1.78		1.77				1.77
R_e(U–O$_{Carb.}$)^b	2.417	2.379	2.423	2.386	2.391	2.366	
	2.416	2.379	2.423	2.386	2.391	2.366	
R_e(U–ON)	2.322	2.310	2.324	2.312	2.391^c	2.366^c	
	2.323	2.310	2.324	2.312	2.391^c	2.366^c	
$\angle$O$_{ax}$–U–O$_{ax}$	180.0	180.0	180.0	180.0	180.0	180.0	180.0
$\angle$O$_{Carb.}$–U–ON^b	66.7	67.3	66.8	67.4	54.8^c	55.4^c	
	66.7	67.3	66.8	67.4	54.8^c	55.4^c	
E	2393	2412	2361	2379	2242	2259	
$\lg\beta$	30		15.25				

^{a}Composition of the EXAFS samples: sha$^-$: 76% [UO$_2$sha$_2$], 12% [UO$_2$sha]$^+$, 12% UO$_2^{2+}$; bha$^-$: 90% [UO$_2$bha$_2$], 8% [UO$_2$bha]$^+$, 2% UO$_2^{2+}$; UO$_2^{2+}$: 100% UO$_2^{2+}$.
^{b}O$_{Carb.}$ is the oxygen atom of the carbonyl group.
^{c}In the case of ba$^-$ the coordinating oxygen atom denoted as ON corresponds to the original hydroxyl group.

where the nitrogen atoms of the ligands are located on the same side, have also bent O–U–O units analogous to the 1:1 complexes (cf. Table C.8).

The LPP gas phase bond lengths and angles are in good agreement with the SPP data, i.e. the deviations amount at most to -0.050 (2.8%), -0.038 Å (1.6%), and $+0.6°$ (1.1%) for R_e(U–O$_{ax}$), R_e(U–O$_{eq}$), and bond angles, respectively. The differences between the U–O$_{ax}$ bond lengths are comparable to the deviation of 0.050 Å (3.0%) found for the DFT/B3LYP calculation of the bare uranyl ion (cf. Sect. 3.1.8.2).

Table 3.21 shows the calculated structure parameters for the complexes [UO$_2$L$_2$OH$_2$] (L=sha, bha, ba) as well as for the solvated uranyl ion [UO$_2$(OH$_2$)$_5$]$^{2+}$. For SPP calculations an increase of the mean U–O$_{ax}$ bond lengths by about 0.031 Å is observed, if the solvated uranyl ion is complexed. However, for LPP calculations a decrease of these bond lengths by ca. 0.016 Å is found. This is in contrast to the experiments, which rather show a bond length increase analogous to the SPP results. Consequently, for [UO$_2$L$_2$OH$_2$] the experimental U–O$_{ax}$ distances are by ca. 0.01 Å over- and by at most 0.04 Å underestimated by SPP and LPP calculations, respectively. The mean U–O$_{eq}$ bond lengths always overestimate the experimental values by at most 0.05 and 0.08 Å for [UO$_2$L$_2$OH$_2$] and [UO$_2$(OH$_2$)$_5$]$^{2+}$, respectively. This overestimation is most likely connected with the shortcomings of the applied level 3 solvation model and the DFT/B3LYP method [158]. In contrast to the gas phase structures the mean U–O$_{eq}$ distances differ only slightly for all complexes, i.e. the deviations amount at most to 0.004 and 0.008 Å between the sha and bha complexes and the sha/bha and ba complexes, respectively. Moreover, bent O–U–O units are obtained, if the solvation effects are included. This is probably due to the fact that the complexes' symmetry is reduced from C_i to C_1.

The LPP and SPP structures including the solvation model deviate at most by -0.049 (2.8%), -0.026 Å (1.1%), and $+1.5°$ (2.3%) for $\overline{R_e}$(U–O$_{ax}$), $\overline{R_e}$(U–O$_{eq}$), and bond angles, respectively. Thus, the solvated structures confirm the applicability of the hexavalent uranium LPP for these complexes.

Relative Stabilities The zero-point corrected complex binding energies (cf. Table 3.20) were obtained by $E = E(\text{UO}_2^{2+}) + 2 \times E(\text{L}^-) - E([\text{UO}_2\text{L}_2])$ and the zero-point energies were scaled by 0.972 [159]. Analogous to the 1:1 complexes only gas phase binding energies are given, because explicit hydration was exclusively considered for the UO$_2^{2+}$ fragment and not for ligands.

Table 3.21: Bond lengths R_e (in Å) and angles $\angle$ (in deg) for the complexes [UO$_2$L$_2$OH$_2$] (L=sha, bha, ba) and the solvated uranyl ion [UO$_2$(OH$_2$)$_5$]$^{2+}$ (given as (UO$_2^{2+}$)$_{aq}$) from SPP and LPP DFT/B3LYP calculations including solvation effects compared to experimental EXAFS data [84]. Additionally, f orbital occupations from Mulliken population analyses are given.

	[UO$_2$sha$_2$OH$_2$]		[UO$_2$bha$_2$OH$_2$]		[UO$_2$ba$_2$OH$_2$]		(UO$_2^{2+}$)$_{aq}$
	SPP	LPP	SPP	LPP	SPP	LPP	SPP
R_e(U–O$_{ax}$)	1.786	1.740	1.781	1.736	1.772	1.728	1.749
	1.782	1.734	1.784	1.736	1.776	1.723	1.749
$\overline{R}_e$(U–O$_{ax}$)	1.784	1.737	1.783	1.736	1.774	1.725	1.749
$R_{exp.}$(U–O$_{ax}$)^a	1.78		1.77				1.77
R_e(U–O$_{Carb.}$)^b	2.454	2.421	2.412	2.427	2.386	2.372	
	2.402	2.373	2.464	2.379	2.438	2.455	
R_e(U–ON)	2.325	2.312	2.436	2.318	2.393^c	2.362^c	
	2.434	2.420	2.326	2.427	2.482^c	2.418^c	
R_e(U–OH$_2$)	2.600	2.563	2.594	2.558	2.554	2.518	
$\overline{R}_e$(U–O$_{eq}$)	2.443	2.418	2.447	2.422	2.451	2.425	2.496
$R_{exp.}$(U–O$_{eq}$)^a	2.42		2.40				2.42
$\angle$O$_{ax}$–U–O$_{ax}$	175.8	176.5	176.0	176.2	178.3	178.4	179.8
$\angle$O$_{Carb.}$–U–ON^b	64.9	65.5	65.2	66.6	54.2^c	54.1^c	
	65.6	66.4	65.8	65.7	53.5^c	54.7^c	
f occ.	2.51	2.06	2.51	2.03	2.54	2.09	

^{a}Composition of the EXAFS samples: sha$^-$: 76% [UO$_2$sha$_2$], 12% [UO$_2$sha]$^+$, 12% UO$_2^{2+}$; bha$^-$: 90% [UO$_2$bha$_2$], 8% [UO$_2$bha]$^+$, 2% UO$_2^{2+}$; UO$_2^{2+}$: 100% UO$_2^{2+}$.
^{b}O$_{Carb.}$ is the oxygen atom of the carbonyl group.
^{c}In the case of ba$^-$ the coordinating oxygen atom denoted as ON corresponds to the original hydroxyl group.

The binding energies show the same trend as the experimental stability constants, i.e. they decrease from [UO$_2$sha$_2$] via [UO$_2$bha$_2$] to [UO$_2$ba$_2$]. While the decrease from the sha to the bha system's SPP binding energies only amounts to 32 kJ mol^{-1}, that between the bha and ba systems is more than three times larger (119 kJ mol^{-1}). A similar observation is made for the 1:1 complexes, i.e. the energy difference between the sha and bha systems amounts to 33 kJ mol^{-1} and that between the bha and ba systems is 106 kJ mol^{-1} (cf. Table 3.18). Thus, the complexes formed with the model ligand ba$^-$ differ clearly from the hydroxamate systems.

Analogous to the molecular structures, the comparison between LPP and SPP binding energies shows an excellent agreement, i.e. they differ at most by 19 kJ mol^{-1} corresponding to 0.8 %. The successful application to these complexes can be explained by the sufficiently small deviations between LPP and SPP 5f occupations of at most 0.48 electrons (cf. Table 3.21). Thus, the uranyl(VI) complexes' structures and binding energies can also be investigated using the hexavalent LPP for uranium, whereby at least some computational time can be saved, i.e. DFT/B3LYP single-point calculations for [UO$_2$sha$_2$OH$_2$] need 1309 and 1129 s using the SPP and LPP, respectively.

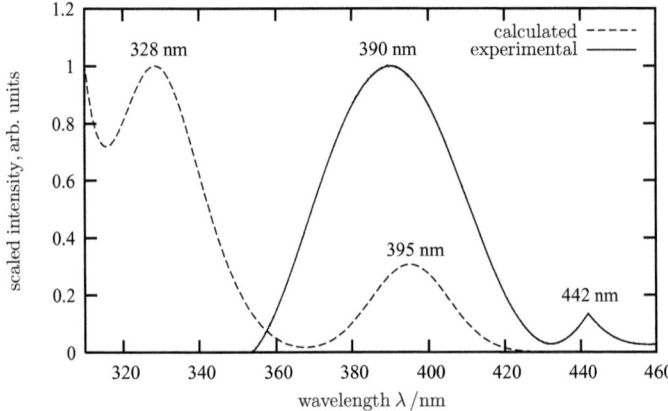

Figure 3.12: Calculated SPP TD-DFT spectrum for the [UO$_2$sha$_2$OH$_2$] complex in comparison to the experimental UV–vis spectrum [126] from aqueous solution. Calculated intensities were scaled to have relative intensities of one for the 328 nm excitation. The small peak at 442 nm is a baseline correction artifact and of no significance.

Electronic Spectra Figure 3.12 shows the calculated SPP TD-DFT spectrum for the [UO$_2$sha$_2$OH$_2$] complex in comparison to the experimental UV–vis spectrum [126]. The experimental spectrum shows one broad peak at 390 nm and the calculated spectrum shows two peaks at 328 and 395 nm. Both calculated excitations are CT transitions from L$^-$ π MOs to U 5f atomic-orbital-like MOs. Since the calculated maximum is only set off by 5 nm from the experimental one, the calculation seems to be in excellent agreement with the experiment.

In the case of the calculated SPP TD-DFT and experimental spectra for [UO$_2$bha$_2$OH$_2$]

3.1 5F-IN-CORE PSEUDOPOTENTIALS FOR ACTINIDES

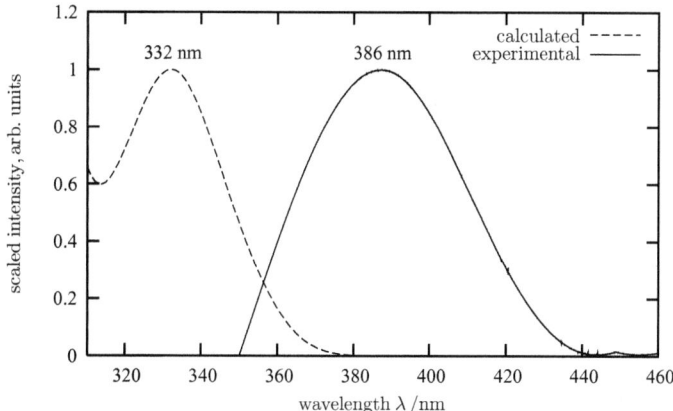

Figure 3.13: Calculated SPP TD-DFT spectrum for the [UO$_2$bha$_2$OH$_2$] complex in comparison to the experimental UV–vis spectrum [126] from aqueous solution. Calculated intensities were scaled to have relative intensities of one for the 332 nm excitation.

one peak is obtained in the 300–500 nm region (cf. Fig. 3.13). However, the SPP peak, which corresponds to a UO$_2^{2+}$ ← L$^-$ CT excitation, is significantly blue-shifted by 54 nm or 0.52 eV with respect to the experimental absorption band. This blue-shift is probably due to the TD-DFT method, which is known to face problems in the description of CT excitations [167]. Since the [UO$_2$sha$_2$OH$_2$] and [UO$_2$bha$_2$OH$_2$] complexes are quite similar, the agreement with the experiment in the case of the sha complex is most likely fortuitous and the TD-DFT method seems not to be suitable to investigate these complexes.

Figure 3.14 shows the TD-DFT spectra for [UO$_2$sha$_2$OH$_2$] from SPP and LPP calculations in comparison to the experimental UV–vis spectrum [126]. Here, the LPP calculation was performed without the g function, i.e. a (7s6p5d2f)/[6s5p4d2f] basis set was used. As one can see the LPP spectrum shows only one peak within the 300–500 nm region, which is located at 308 nm. However, this peak does not involve a UO$_2^{2+}$ ← L$^-$ CT, but a L$^-$ π^* ← π excitation. Thus, the LPP spectrum even does not show the experimental CT transition. This is most likely due to the fact that the CT excitations from L$^-$ π MOs to U 5f atomic-orbital-like MOs cannot be described by the 5f-in-core LPP, because it cannot accommodate additional electrons in the 5f

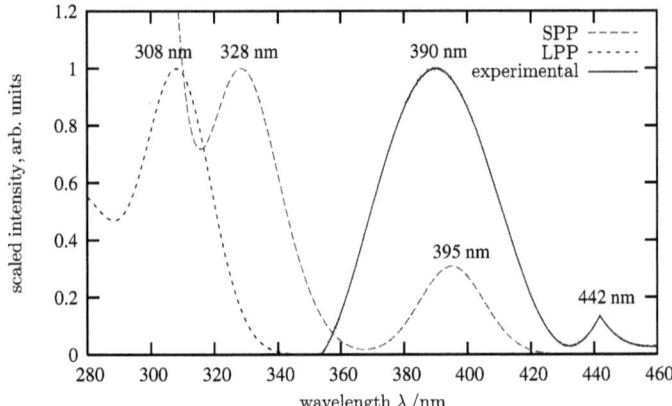

Figure 3.14: TD-DFT spectra for the [UO$_2$sha$_2$OH$_2$] complex from SPP and LPP calculations in comparison to the experimental UV–vis spectrum [126] from aqueous solution. Calculated intensities were scaled to have relative intensities of one for the 308 and 328 excitations, respectively. The small peak at 442 nm is a baseline correction artifact and of no significance.

shell, which is included in the core. Although the f-part of the LPP allows for some additional 5f occupation, it is already exhausted due to the ligand-to-metal donation of more than 2 electrons (cf. 5f occupations in Table 3.21).

3.2 4f-in-core Pseudopotentials for Lanthanides

In this Section the adjustment of the quasirelativistic energy-consistent 4f-in-core PPs for tetravalent lanthanides [28] and of corresponding valence basis sets will be described. Furthermore, molecular test calculations using these PPs and basis sets as well as using the recently published valence basis sets [168] for trivalent 4f-in-core PPs [17] will be discussed, in order to demonstrate the transferability of the PPs and basis sets to a molecular environment.

3.2.1 Adjustment of the Pseudopotentials

The 4f-in-core PPs[8] corresponding to tetravalent lanthanide atoms ($4f^{n-1}$, n=1–3, 8, 9 for Ce–Nd, Tb, Dy) were generated analogous to the quasirelativistic di- ($4f^{n+1}$, n=0–13 for La–Yb) and trivalent ($4f^n$, n=0–14 for La–Lu) 4f-in-core PPs [17, 18]. The parameters for the tetravalent PPs are listed in Table A.3.

The 1s–4f (spherically averaged) shells are included in the PP core, while all orbitals with main quantum number larger than four are treated explicitly, i.e. 12 valence electrons. The s-, p-, and d-PPs are composed of two Gaussians each and were adjusted by a least-squares fit to the total valence energies of 18 reference states (cf. Table A.1). The reference data were taken from relativistic AE calculations using the WB approach. Both AE WB and PP calculations were performed with an atomic finite-difference HF scheme [55].

In order to allow for some participation of the 4f orbitals in chemical bonding the f-parts of the PPs are designed to describe partial occupations of the 4f shell, which are larger than the integral occupation number implied by the valency, i.e. $4f^{n-1+q}$ (n=1–3, 8, 9 for Ce–Nd, Tb, Dy) with $0 \leq q < 1$ for tetravalent lanthanide atoms [18]. In slight variation to the former PPs [18] and analogous to the 5f-in-core PPs for actinides [20], the f-PPs consist of two types of potentials V_1 and V_2 (cf. (3.1)).

The errors in the total valence energies of finite-difference HF calculations are smaller than 0.07 and 0.21 eV (0.4%) for s-, p-, d- and f-parts, respectively (cf. Tables A.5 and A.9).

---

[8]The tetravalent 4f-in-core PPs were adjusted by M. Hülsen.

3.2.2 Optimization of the Valence Basis Sets

In the following the optimization of the valence basis sets[9] corresponding to tetravalent 4f-in-core PPs [28] will be presented in two parts, i.e. first the generation of the primitive and then that of the segmented contracted basis sets will be given. The basis set parameters are listed in Table B.11.

3.2.2.1 Primitive Basis Sets

The GTO valence basis sets were constructed analogous to those for tetravalent 5f-in-core PPs for actinides [29]. First, basis sets for use in crystal calculations were created, i.e. (4s4p3d) and (5s5p4d) basis sets were HF energy-optimized [57] for the valence subconfiguration Ln^{3+} $5s^25p^65d^1$. All exponents, which became smaller than 0.15, were fixed to this value and the remaining exponents were reoptimized. Furthermore, all optimizations were carried out with the requirement that the ratio of exponents in the same angular symmetry must be at least 1.5. The basis set errors in the valence energies with respect to numerical finite-difference 4f-in-core LPP HF calculations [55] are below 0.15 and 0.03 eV for (4s4p3d) and (5s5p4d), respectively (cf. Table B.12). Secondly, the valence basis sets were augmented by adding a set of 2s1p1d low-exponent Gaussians yielding (6s5p4d) and (7s6p5d) primitive sets for use in molecular calculations. The added exponents were HF energy-optimized [57] for the $5s^25p^65d^26s^2$ valence subconfiguration. The differences in the valence energies with respect to numerical finite-difference LPP HF calculations [55] are below 0.15 and 0.03 eV for (6s5p4d) and (7s6p5d), respectively (cf. Table B.12). Finally, sets of 2f1g polarization functions were energy-optimized in CI calculations [52] for the $5d^26s^2$ valence subconfiguration.

Table 3.22: Contraction schemes.

Contraction	Contraction Scheme	
	(6s5p4d)	(7s6p5d)
[4s3p3d]	{3111/311/211}	{3211/411/311}
[5s4p4d][a]	{21111/2111/211}	{31111/2211/2111}
[6s5p4d]		{211111/21111/2111}

[a]In the case of the (6s5p4d) basis set the VTZ contraction is [5s4p3d].

---
[9]The valence basis sets were optimized by M. Hülsen.

3.2.2.2 Contracted Basis Sets

The basis sets were contracted using different segmented contraction schemes (cf. Table 3.22) to yield basis sets of approximately VDZ, VTZ, and VQZ quality for the s and p symmetries. In the case of d symmetry at least a triple-zeta contraction was necessary and additional sets with a less tight d contraction are also offered (VDZ: [4s3p3d], VTZ: [5s4p3d], [5s4p4d], and VQZ: [6s5p4d]).

The errors in total valence energies of the $5d^2 6s^2$ valence substates with respect to numerical finite-difference LPP HF calculations [55] are listed in Table B.13. For the VDZ and VTZ contractions of the (6s5p4d) basis sets these errors are below 0.21 and 0.17 eV, respectively. For the VDZ, VTZ, and VQZ contractions of the (7s6p5d) basis sets these errors are below 0.07, 0.05, and 0.03 eV, respectively.

3.2.3 Molecular Test Calculations

For the tetravalent 4f-in-core LPPs and corresponding valence basis sets molecular test calculations were performed for lanthanide tetrafluorides LnF_4 and cerium dioxide CeO_2, respectively [28]. Furthermore, the recently published valence basis sets [168] for trivalent 4f-in-core LPPs [17] were tested by calculating lanthanide trifluorides LnF_3 [54]. The calculations for LnF_4, CeO_2, and LnF_3 will be discussed separately. Finally, the results for LnF_3 will be compared to those for AnF_3.

3.2.3.1 Lanthanide Tetrafluorides

The LPP HF and CCSD(T) calculations for LnF_4 (Ln=Ce–Nd, Tb, Dy) and CeF_4 will be compared to corresponding SPP state-averaged MCSCF calculations and experimental [169] as well as computational [170] data from the literature, respectively. The results for bond lengths as well as bond energies and those of Mulliken population analyses are listed in Tables 3.23 and 3.24, respectively.

Computational Details The test calculations[10] for LnF_4 (Ln=Ce–Nd, Tb, Dy) were carried out with the MOLPRO program package [52] implying T_d symmetry and using tetravalent 4f-in-core LPPs as well as 4f-in-valence SPPs [14]. For F Dunning's aug-cc-pVQZ basis set [62, 63] was applied, and for Ln (7s6p5d2f1g)/[6s5p4d2f1g] and (14s13p10d8f6g)/[6s6p5d4f3g] [171] valence basis sets were used for LPP HF and

---

[10] The calculations for LnF_4 were performed by M. Hülsen.

Table 3.23: Ln–F bond lengths R_e (in Å) and energies E_{bond} (in eV) for LnF$_4$ (Ln=Ce–Nd, Tb, Dy) from LPP HF and SPP state-averaged MCSCF calculations. For CeF$_4$ LPP CCSD(T) as well as experimental [169] data are given in italics.

Ln	R_e		E_{bond}^a	
	LPP	SPP[b]	LPP	SPP
Ce	2.045	2.031	5.269	5.386
	2.048	*2.036(5)*	*6.715*	
Pr	2.033	2.017	5.236	5.291
Nd	2.021	2.005	5.219	5.222
Tb	1.963	1.957	5.245	4.989
Dy	1.952	1.946	5.265	5.043

[a]Bond energies are not corrected to account for the experimentally observed ground states.
[b]Given in italics: experimental bond length for CeF$_4$.

SPP state-averaged MCSCF calculations, respectively. The state-averaging was necessary to avoid symmetry-breaking at the orbital level, because the program MOLPRO is limited to the D_{2h} point group and subgroups. For CeF$_4$ a LPP CCSD(T) calculation was performed, since for this compound an experimental bond length [169] is available. In the CCSD(T) calculation for F Dunning's aug-cc-pVTZ basis set [62, 63] was applied, and the F 1s orbitals were frozen.

The Ln–F bond energy for LnF$_4$ was calculated by $E_{bond} = [E(Ln) + 4 \times E(F) - E(LnF_4)]/4$, where the lanthanide atom was assumed to be in the lowest valence substate, i.e. $4f^{n-1}5d^26s^2$. In order to calculate bond energies with respect to the experimentally observed ground states of the lanthanides, the strategy proposed for the di- and trivalent 4f-in-core PPs almost two decades ago [70] should be used (cf. Sect. 3.1.5.1). In contrast to di- and trivalent for tetravalent 4f-in-core PPs an energy correction using experimental energy differences is not possible, since for the $5d^26s^2$ valence subconfiguration no experimental data are available [71]. If desired, correlation contributions can be obtained by SPP or AE atomic calculations. Tables C.9 and C.10 summarize AE WB and AE DHF corrections, respectively.

Lanthanide Tetrafluoride Structure Due to the lanthanide contraction the Ln–F bond lengths calculated by using LPP HF and SPP state-averaged MCSCF decrease continuously with increasing nuclear charge by 0.093 and 0.085 Å, respectively. The LPP HF bond lengths are in good agreement with the SPP reference data, i.e. the

m.a.e. and m.r.e. amount to 0.012 Å and 0.6%, respectively. The maximum error is 0.016 Å (0.8%). Compared to the Ce–F bond length of 2.036 Å [170] calculated by using an ECP (core: 1s–4d) at the HF level, the LPP HF value also deviates only by 0.009 Å corresponding to 0.4%.
If correlation is included via CCSD(T), the bond length of CeF_4 becomes slightly longer by 0.003 Å. The difference between the LPP CCSD(T) and experimental [169] Ce–F bond lengths amount to 0.012 Å (0.6%). The deviation to the Ce–F bond length of 2.041 Å from an ECP MP2 calculation [170] is 0.007 Å (0.3%). Thus, the LPP CCSD(T) bond length of CeF_4 is also in good agreement with corresponding reference data and confirms the reliability of the newly developed LPPs.

Lanthanide–fluorine Bond Energy While the LPP HF Ln–F bond energies (for the lowest valence substates of the superconfigurations) stay almost constant ($\Delta E_{max}=$ 0.05 eV), the SPP state-averaged MCSCF bond energies (for the lowest states of the configuration) decrease with increasing nuclear charge and show a minimum for Tb ($\Delta E_{max}=0.40$ eV), i.e. for the half-filled 4f shell. The LPP and SPP bond energies show a satisfactory agreement, i.e. the m.a.e. (m.r.e.) amounts to 0.13 eV (2.6%) and the maximum error, which occurs for Tb, is 0.26 eV (5.1%). For CeF_4 the LPP HF bond energy is by 0.34 eV (6.9%) larger than the value obtained by Lanza and Fragala in an ECP HF calculation (4.93 eV) [170]. However, this is most likely due to the different basis sets rather than to the different core definitions (basis sets: LPP: Ce (7s6p5d2f1g)/[6s5p4d2f1g], F (13s7p4d3f2g)/[6s5p4d3f2g]; ECP: Ce [4s4p2d2f], F (11s6p2d)/[5s3p2d]; core: LPP: 1s–4f; ECP: 1s–4d).
As expected the inclusion of electron correlation via CCSD(T) clearly increases the Ce–F bond energy by 1.45 eV. The LPP CCSD(T) bond energy agrees quite well with the ECP MP2 bond energy of 6.73 eV obtained by Lanza and Fragala [170], i.e. the difference amounts to 0.015 eV (0.2%).

Mulliken Orbital Populations The Mulliken orbital populations show that the bonding in LnF_4 is basically ionic with significant back-bonding into the Ln 5d and 4f (less 6s) orbitals. For LPP and SPP calculations this results in charge separations up to 0.84 and 0.72 electrons per bond and in total atomic charges of up to 3.35 and 2.87 units on the lanthanide, respectively. The SPP 4f occupations deviate on average by 0.27 and at most by 0.35 electrons from the assumed LPP $4f^{n-1}$ occupations. This demonstrates that the 4f orbitals participate to some extent in the Ln–F bonding. However, the 4f-in-core approach yields reasonable results, since the mean differences between LPP and SPP 4f occupations amount only to 0.13 and at most to 0.19 electrons, because the f-part of the LPPs allows for some 4f occupation in addition to the integral $4f^{n-1}$ assumption.

Table 3.24: Mulliken 5s/6s, 5p, 5d, and 4f orbital populations and atomic charges (Q) on Ln in LnF$_4$ (Ln=Ce–Nd, Tb, Dy) from LPP HF and SPP state-averaged MCSCF calculations. A $5s^2 5p^6 5d^2 6s^2$ ground state valence subconfiguration is considered for Ln.

Ln	s		p		d		f		Q	
	LPP	SPP	LPP	SPP	LPP	SPP	LPP[a]	SPP	LPP	SPP
Ce	1.98	2.13	5.86	5.91	0.69	0.79	0.16	0.35	3.30	2.78
Pr	1.97	2.14	5.86	5.90	0.70	0.81	0.15	1.33	3.31	2.77
Nd	1.97	2.15	5.85	5.90	0.71	0.80	0.14	2.30	3.32	2.80
Tb	1.97	2.23	5.84	5.88	0.73	0.82	0.12	7.18	3.34	2.87
Dy	1.97	2.25	5.83	5.87	0.73	0.90	0.11	8.17	3.35	2.79

[a] 0, 1, 2, 7, and 8 electrons in the 4f shell are attributed to the LPP core for Ce, Pr, Nd, Tb, and Dy, respectively.

3.2.3.2 Cerium Dioxide

The HF and CCSD(T) calculations for CeO$_2$ using the tetravalent LPP will be compared to 4f-in-valence SPP reference data [172]. The results for bond lengths, bond angles, binding energies, and f orbital populations are listed in Table 3.25.

Computational Details The LPP HF, LPP CCSD(T), and SPP [14] HF optimizations for CeO$_2$ were performed with MOLPRO [52] using C_{2v} symmetry. For Ce (7s6p5d2f1g)/[6s5p4d2f1g] and (14s13p10d8f6g)/[6s6p5d4f3g] [171] valence basis sets were used for LPP and SPP calculations, respectively, and for O Dunning's aug-cc-pVQZ basis set [62, 63] was applied. In the case of the CCSD(T) calculation the O 1s orbitals were frozen, and for O Dunning's aug-cc-pVTZ basis set [62, 63] was used. The binding energy of CeO$_2$ was calculated by $E_{bond} = E(Ce) + 2 \times E(O) - E(CeO_2)$, where the cerium atom was assumed to have the $4f^0 5d^2 6s^2$ valence subconfiguration.

Hartree–Fock Results The LPP HF molecular structure for CeO$_2$ deviates significantly from the SPP HF reference data, i.e. the differences between the Ce–O bond

3.2 4F-IN-CORE PSEUDOPOTENTIALS FOR LANTHANIDES

Table 3.25: Ce–O bond lengths R_e (in Å), bond angles $\angle$O–Ce–O (in deg), binding energies E_{bond} (in eV), and Mulliken f orbital populations for CeO_2 from LPP HF, LPP CCSD(T), and SPP HF calculations. Furthermore, results of a HF calculation using the old tetravalent LPP [173] as well as SPP data [172] from the literature are given.

Method	R_e	$\angle^a$	E^b_{bond}	f occ.
LPP HF	1.839	107.6	9.23	0.18
LPP HF^c	1.782	118.4	11.61	0.50
LPP HF^d	1.886	106.8	8.09	0.10
old LPP HF	1.892	107.1	7.98	0.08
SPP HF	1.790	116.7	11.07	0.67
SPP HF [172]	1.792	118.1		0.69
LPP CCSD(T)	1.877	104.2	15.36	0.19
LPP CCSD(T)^c	1.818	124.4	18.65	0.51
SPP CISD+Q [172]	1.804	118.8		
SPP ACPF [172]	1.838	117.6		

^{a}Experimental bond angle: $\angle$O–Ce–O=146±2 [174].
^{b}Binding energies are not corrected to account for the experimentally observed ground states.
^{c}LPP HF calculation using a (7s6p5d3f1g)/[6s5p4d3f1g] basis set for Ce, where the 3f basis functions were HF optimized for $4f^25s^25p^66s^2$.
^{d}LPP HF calculation using a V_2 potential as f-projector.

lengths and O–Ce–O bond angles amount to 0.049 Å (2.7%) and 9.1° (7.8%), respectively. Compared to the SPP HF data from the literature [172] these deviations are 0.047 Å (2.6%) and 10.5° (8.9%). In the case of the binding energies the deviation between LPP and SPP data is even larger and amounts to 1.84 eV (16.6%). The reason for these significant discrepancies is the large deviation between the LPP and SPP f occupations, which amounts to 0.49 electrons.

The HF results using the old tetravalent LPP, which was adjusted in 1991 to calculate cerocene [173], deviate even more from the SPP data, i.e. the differences are 0.102 Å (5.7%), 9.6° (8.2%), 3.09 eV (27.9%), and 0.59 electrons for bond lengths, bond angles, binding energies, and f occupations, respectively. These larger deviations are due to the fact that the f-projector of the old LPP does not allow for any 4f participation ($V = V_2$ in (3.1)). Therefore the f occupation is very small (0.08 electrons) and corresponds to the occupation of the 5f, 6f, ... shells.

If a V_2 instead of a V_1 potential is used as f-projector for the new LPP, the results become very similar to those using the old LPP, and the remaining deviations can be explained by the different basis sets, i.e. the deviations amount to 0.006 Å, 0.3°,

0.11 eV, and 0.02 electrons. Thus, the use of a f-projector, which admits some 4f occupation, is important especially for Ce, where the 4f shell is unoccupied. However, with increasing 4f occupation along the lanthanide series this additional occupation should get less probable, wherefore the ratio of V_2 is increased continuously by $1/14$ with increasing 4f occupation (cf. (3.1)). The differences between results for $V = V_1$ and $V = V_2$ are 0.047 Å, 0.8°, 1.14 eV, and 0.08 electrons, respectively. The influence of the mixing ratio of V_1 and V_2 within (3.1) should be smaller than these deviations. In order to improve the LPP results, it was tried to use 3f basis functions HF optimized [52] for the valence configuration $4f^2 5s^2 5p^6 6s^2$ instead of 2f polarization functions CI optimized [52] for $5s^2 5p^6 5d^2 6s^2$ (exponents: 3f: 8.4453, 2.7912, 0.7481; 2f: 0.9916, 0.3239). The new results are in good agreement with the SPP HF data, i.e. the deviations amount to 0.008 Å (0.4%), 1.7° (1.5%), 0.54 eV (4.9%), and 0.17 electrons, respectively. Moreover, the molecular energy E(CeO$_2$) is reduced by 2.39 eV (-187.975 vs. -187.887 a.u.), which shows that the modified basis set performs clearly better. The reason for this better performance is probably that the 3f basis functions are not as diffuse as the 2f polarization functions and therefore allow for more additional 4f occupation. If the 3f basis functions are applied for CeF$_4$, the deviations from the SPP bond length, bond energy, and f occupation are slightly reduced to 0.004 Å, 0.09 eV, and 0.17 electrons, respectively (deviations for 2f: 0.014 Å, 0.12 eV, 0.19 electrons). Furthermore, the molecular energy E(CeF$_4$) is reduced by 0.82 eV (-436.365 vs. -436.335 a.u.).

In the case of the other lanthanides Pr, Nd, Tb, and Dy f exponents HF optimized [52] for $4f^2 5s^2 5p^6 6s^2$ amount at most to 0.3140 and are thus more diffuse than the f polarization functions, where the smallest exponent is 0.3270. This is most likely due to the admixture of the V_2 potential in the f-PP (cf. (3.1)), which does not allow for an additional 4f occupation. For these elements LPP HF calculations for LnF$_4$ using the 3f basis functions give by 0.17–0.26 eV higher molecular energies E(LnF$_4$) indicating that these basis sets are not as good as the 2f polarization functions. Furthermore, the deviations between LPP and SPP bond lengths are increased, if the 3f basis functions are used (Pr 0.024 vs. 0.016, Nd 0.022 vs. 0.015, Tb 0.010 vs. 0.006, Dy 0.010 vs. 0.007 Å). Thus, in the case of these elements the 2f polarization functions should be used. However, if the 3f basis functions are used for CeF$_4$ and the 2f polarization functions are used for Pr, Nd, Tb, and Dy, the bond lengths show no regular variation with increasing nuclear charge, but a skip of 0.007 Å between CeF$_4$ and PrF$_4$ (Ce 2.026, Pr

2.033, Nd 2.021, Tb 1.963, Dy 1.952 Å).

Coupled Cluster Results If CCSD(T) instead of HF calculations are carried out using the 2f polarization functions, the Ce–O bond length is increased by 0.038 Å, the O–Ce–O bond angle is decreased by 3.4°, the binding energy is increased by 6.13 eV, and the f occupation stays almost constant (Δf occ.=0.01 electrons). The deviations from SPP structures determined by the CISD method including the correction formula proposed by Langhoff and Davidson (CISD+Q) and by the ACPF method [172] are quite large and amount to 0.073 (4.0%) and 0.039 Å (2.1%) and 14.6 (12.3%) and 13.4° (11.4%) for bond lengths and angles, respectively. The deviation from the experimental bond angle of 146±2° [174] is even larger and amounts to 42° (29%). However, this deviation can partly be explained by the fact that the experimental value determined based on the infrared spectrum of CeO_2 in an Ar matrix does not include corrections for anharmonicity effects (estimated to reduce the bond angle by 5–10°) and the influence of the matrix on the bond angle.

Using the 3f basis functions instead of the 2f polarization functions yields clearly smaller differences from SPP CISD+Q and ACPF data [172], respectively, i.e. the deviations are reduced to 0.014 (0.8%) and 0.020 Å (1.1%) and 5.6 (4.7%) and 6.8° (5.8%) for bond lengths and angles, respectively. Compared to the experimental bond angle the deviation amounts to 22° (15%), which is by about 50% smaller than that using the 2f polarization functions. Thus, also at the correlated level the use of the 3f basis functions adjusted to $4f^25s^25p^66s^2$ shows a considerable improvement.

3.2.3.3 Lanthanide Trifluorides

The LnF_3 HF and CCSD(T) results calculated using 4f-in-core LPPs [17] with and without CPPs [41] and the recently published (8s7p6d3f2g)/[6s5p5d3f2g] basis sets [168] will be compared to SPP [14] state-averaged MCSCF and experimental [175–177] data from the literature, respectively (cf. Tables 3.26/3.27 for HF and 3.28/3.29 for CCSD(T) results, respectively) [54]. However, the experimental atomization energies [176] were not measured directly, but calculated from a thermochemical cycle, where sometimes estimated values were used. The assumed uncertainty is ±0.22 eV if all quantities are measured (La, Pr, Nd, Gd, Er) and ±0.43 eV if one or more estimated values are used (Ce, Sm, Eu, Tb–Ho, Tm–Lu).

In the case of Gd the contraction coefficients of the s basis functions were modified,

Table 3.26: Ln–F bond lengths R_e (in Å) and bond angles $\angle$F–Ln–F (in deg) for LnF$_3$ (Ln=La–Lu) from HF calculations using LPPs with and without CPPs in comparison to SPP state-averaged MCSCF and LPP HF [178] calculations.

Ln	R_e				$\angle$F–Ln–F			
	LPP	CPP^a	SPP	[178]	LPP	CPP^a	SPP	[178]
La	2.146	2.143	2.138	2.15	118.8	118.7	119.5	116.0
Ce	2.132	2.128	2.119	2.13	119.4	119.3	118.4	116.8
Pr	2.118	2.113	2.107	2.12	119.7	119.7	118.8	117.4
Nd	2.105	2.098	2.096	2.11	119.9	119.9	119.5	118.0
Pm	2.092	2.085	2.086	2.09	120.0	120.0	120.0	118.9
Sm	2.080	2.071	2.075	2.08	120.0	120.0	120.0	119.8
Eu	2.067	2.057	2.065	2.06	120.0	120.0	120.0	120.0
Gd	2.056	2.045	2.054	2.06	120.0	120.0	120.0	120.0
Tb	2.043	2.031	2.043	2.05	120.0	120.0	120.0	120.0
Dy	2.032	2.018	2.032	2.03	120.0	120.0	120.0	120.0
Ho	2.020	2.005	2.021	2.02	120.0	120.0	120.0	120.0
Er	2.009	1.992	2.010	2.01	120.0	120.0	120.0	120.0
Tm	1.999	1.980	2.000	2.00	120.0	120.0	120.0	120.0
Yb	1.988	1.968	1.989	1.99	120.0	120.0	120.0	120.0
Lu	1.984	1.963	1.978	1.98	120.0	120.0	120.0	120.0

^{a}LPP calculations using CPPs.

because new coefficients reduce the energy of Gd^{3+} by more than 1 a.u. (s coefficients: old -0.9361, 1.3080, -0.8824; new -0.0801, 0.5277, -0.9570).

Computational Details For LnF$_3$ (Ln=La–Lu) HF and CCSD(T) calculations using trivalent 4f-in-core LPPs [17] with and without CPPs [41] were carried out with MOLPRO [52] implying C_{3v} symmetry. The recently published (8s7p6d)/[6s5p5d] basis sets [168] with newly optimized 3f2g polarization functions[11] were used for Ln. These polarization functions were energy-optimized in MRCI calculations. For F Dunning's aug-cc-pVQZ basis set [62, 63] was applied, and the F 1s orbitals were frozen at the CCSD(T) level. As comparison SPP [14] state-averaged MCSCF calculations were performed [52] using (14s13p10d8f6g)/[6s6p5d4f3g] basis sets [171].
For LaF$_3$ and LuF$_3$ SPP CCSD(T) calculations and for LuF$_3$ an AE/DKH CCSD(T) calculation[12] were carried out [52] using C_{3v} symmetry and Dunning's aug-cc-pVQZ

---
[11] The polarization functions were optimized by J. Yang.
[12] The AE calculation for LuF$_3$ was performed by M. Dolg.

3.2 4F-IN-CORE PSEUDOPOTENTIALS FOR LANTHANIDES

Table 3.27: Ionic binding energies ΔE_{ion} (in eV) for LnF$_3$ (Ln=La–Lu) from HF calculations using LPPs with and without CPPs in comparison to SPP state-averaged MCSCF calculations. Additionally, LPP and SPP f orbital occupations from Mulliken population analyses are given.

Ln	ΔE_{ion}			f occ.	
	LPP	CPP[a]	SPP	LPP[b]	SPP
La	44.48	44.60	44.65	0.10	0.17
Ce	44.83	44.98	45.08	0.10	1.16
Pr	45.17	45.35	45.43	0.09	2.15
Nd	45.50	45.72	45.72	0.09	3.13
Pm	45.81	46.07	46.00	0.09	4.12
Sm	46.11	46.41	46.28	0.09	5.11
Eu	46.42	46.77	46.53	0.09	6.10
Gd	46.69	47.09	46.81	0.09	7.10
Tb	47.00	47.45	47.06	0.09	8.09
Dy	47.29	47.80	47.36	0.09	9.09
Ho	47.58	48.16	47.64	0.09	10.08
Er	47.87	48.51	47.92	0.09	11.07
Tm	48.14	48.85	48.19	0.09	12.07
Yb	48.43	49.21	48.46	0.09	13.06
Lu	48.58	49.44	48.80	0.08	14.04

[a] LPP calculations using CPPs.
[b] 0–14 electrons in the 4f shell are attributed to the LPP core for La–Lu, respectively.

basis set for F [62, 63]. In the SPP calculations for Ln (14s13p10d8f6g)/[6s6p5d4f3g] basis sets [171] were used and the Ln 4s, 4p, and 4d as well as the F 1s orbitals were frozen. In the AE calculation for Lu a (26s22p18d14f6g)/[10s8p6d5f3g] generalized contracted basis set was applied and the Lu 1s–4s, 2p–4p, and 3d–4d as well as the F 1s orbitals were frozen.

Lanthanide Trifluoride Structure The LPP HF bond lengths calculated using the new basis sets [168] overestimate the SPP bond lengths on average by 0.005 Å (0.2%). If CPPs are applied, the LPP bond lengths decrease by about 0.012 Å. Since this decrease grows along the lanthanide series (ΔR_e: La -0.003; Lu -0.021 Å), the LPP+CPP values still overestimate the SPP bond lengths for the lighter (La–Nd) and underestimate them for the heavier lanthanides (Pm–Lu). Altogether the CPPs increase the deviations from the SPP data, i.e. the m.a.e. (m.r.e.) amounts to 0.011 Å (0.5%). However, it should be noted here that the CPP models static as well as dynamic

Table 3.28: Ln–F bond lengths R_e (in Å) and bond angles $\angle$F–Ln–F (in deg) for LnF$_3$ (Ln=La–Lu) from CCSD(T) calculations using LPPs with and without CPPs in comparison to experimental data [175] as well as to LPP MP2 [178] calculations.

Ln	R_e				$\angle$F–Ln–F^a		
	LPP	CPP^b	Exp.^{c}	[178]	LPP	CPP^b	[178]
La^d	2.133	2.130	*2.139*	2.15	115.8	115.8	112.9
Ce	2.119	2.116	*2.127*	2.13	116.3	116.2	113.7
Pr	2.105	2.101	2.091	2.12	116.5	116.5	114.1
Nd	2.092	2.088	*2.090*	2.10	116.8	116.7	114.6
Pm	2.080	2.076	*2.077*	2.09	117.4	117.2	115.5
Sm	2.069	2.063	*2.065*	2.08	117.9	117.7	116.3
Eu	2.058	2.051	*2.054*	2.06	118.5	118.2	118.3
Gd	2.048	2.041	2.053	2.06	119.1	118.8	117.8
Tb	2.037	2.029	*2.030*	2.05	119.7	119.5	119.1
Dy	2.026	2.017	*2.019*	2.04	120.0	119.8	120.0
Ho	2.015	2.005	2.007	2.02	120.0	120.0	120.0
Er	2.004	1.993	*1.997*	2.01	120.0	120.0	120.0
Tm	1.994	1.982	*1.987*	2.00	120.0	120.0	120.0
Yb	1.984	1.970	*1.975*	1.99	120.0	120.0	120.0
Lu^d	1.982	1.967	*1.968*	1.98	120.0	120.0	120.0

^{a}Experimental $\angle$F–Ln–F (in deg): Pr 105.0±1.5, Gd 109.0±2.3, Ho 110.8±1.2 [175].
^{b}LPP calculations using CPPs.
^{c}Numbers in italics represent estimated values.
^{d}SPP CCSD(T): La R_e=2.121 Å, $\angle$F–La–F=116.2°; Lu R_e=1.962 Å, $\angle$F–Lu–F=120.0°; AE/DKH CCSD(T): Lu R_e=1.967 Å, $\angle$F–Lu–F=120.0°.

core-polarization, whereas the SPP state-averaged MCSCF calculations only account for the former. The bond lengths from LPP HF calculations [178] using the original (7s6p5d)/[5s4p3d] basis sets [17] also differ by about 0.005 Å (0.3%) from the SPP data. The differences between these and the LPP HF bond lengths using the new basis sets are negligible, i.e. they amount to ca. 0.003 Å (0.1%), which is in agreement with the finding of other authors for DyCl$_3$ [179], where the original basis set also gives by 0.001–0.003 Å longer bond lengths than the new one.

The planarity or non-planarity of LnF$_3$ has been the subject of some controversy. However, both experimental and theoretical evidence point to C_{3v} structures for the majority of LnF$_3$ [175, 178, 180]. Here, the LPP HF and SPP state-averaged MCSCF calculations yield C_{3v} symmetry only in the beginning of the lanthanide series, i.e. for LaF$_3$–NdF$_3$ (four elements) and PmF$_3$–LuF$_3$ (11 elements) C_{3v} and D_{3h} symmetry is obtained, respectively. The differences between the LPP and SPP bond angles are quite small, i.e. the m.a.e. and m.r.e. are 0.2° and 0.2%, respectively. The application of CPPs has nearly no effect ($\Delta\angle_{max}$=0.05° (0.04%)), wherefore the differences between LPP+CPP and SPP bond angles are the same as those of pure LPPs. The bond angles of the former LPP HF calculations [178] using the original basis sets deviate slightly more from the SPP results, i.e. the m.a.e. (m.r.e.) amounts to 0.6° (0.5%). Here, C_{3v} symmetry is found for LaF_3–SmF_3 (six elements). Analogous to the bond lengths, the deviations between these former and the LPP HF bond angles using the new basis sets are very small, i.e. the differences are on average 0.7° (0.6%).

If correlation effects are included via CCSD(T), the LPP HF bond lengths are shortened by 0.008 Å (0.4%). The differences from the experimental bond lengths [175] are quite small, i.e. the m.a.e. (m.r.e.) amounts to 0.007 Å (0.3%). The use of CPPs at the CCSD(T) level shows a smaller bond length contraction than for the LPP HF calculations, i.e. the reduction is 0.008 (0.4%) instead of 0.012 Å (0.6%). Analogous to the LPP CCSD(T) An–F bond lengths (cf. Sect. 3.1.5.6), the reason for this smaller contraction is most likely the increased F–F repulsion due to the bond length reduction by the consideration of valence correlation effects. Thus, the improvement of the LPP CCSD(T) results by using CPPs is only ca. 0.002 Å (0.1%), i.e. the m.a.e. (m.r.e.) of the LPP+CPP CCSD(T) bond lengths amounts to 0.005 Å (0.2%). The bond lengths from former LPP calculations using the original basis sets and MP2 [178] differ by ca. 0.013 Å (0.7%) from the experimental values. Thus, these deviations are slightly larger than those of the LPP CCSD(T) results. However, with regard to the different methods and basis sets the deviations between the LPP MP2 and CCSD(T) bond lengths are quite small, i.e. they amount at most to 0.017 Å (0.8%). For LaF_3 and LuF_3 the deviations of the LPP CCSD(T) bond lengths from AE DFT/PBE optimizations using ZORA [181] are also quite small, i.e. the La–F and Lu–F bond lengths are overestimated by 0.022 (1.0%) and 0.013 Å (0.7%), respectively (AE bond lengths: LaF_3 2.111, LuF_3 1.969 Å).

In the case of bond angles the inclusion of correlation effects via CCSD(T) results in a decrease of about 1.3° (1.1%). Due to this decrease, for LaF_3–TbF_3 (nine elements) C_{3v} symmetry is obtained. The mean deviation from experimental data [175] amounts to 10.3° (9.5%). The application of CPPs at the CCSD(T) level gives a larger bond angle decrease compared to the HF level, i.e. the bond angles are reduced by 0.12° (0.1%). Because of this decrease, at the LPP+CPP CCSD(T) level also for DyF_3 C_{3v}

symmetry is obtained and the mean deviation from experimental data is slightly reduced to 10.2° (9.4%). The former LPP MP2 bond angles [178] differ even slightly less from the experimental data, i.e. the m.a.e. (m.r.e.) amounts to 9.0° (8.3%). However, both LPP MP2 and CCSD(T) calculations yield C_{3v} symmetry for the same compounds (LaF$_3$–TbF$_3$; nine elements) and deviate only by about 1.1° (0.9%).

One reason for the large deviations (ca. 10%) between computational and experimental bond angles is possibly that the experimental structures were determined according to electron diffraction data without anharmonicity corrections [175]. However, the anharmonicity in angle bending vibration may seriously affect the angle parameters, i.e. the conclusion on equilibrium molecular geometries cannot be considered fully unambiguous [182]. Furthermore, the LPP CCSD(T) bond angles of LaF$_3$ and LuF$_3$ are in good agreement with SPP CCSD(T) data (cf. Table 3.28), i.e. the differences amount to 0.4 and 0.0° for LaF$_3$ and LuF$_3$, respectively.

Compared to the AE DFT/PBE bond angle of LaF$_3$ ($\angle$=113.6°) [181] the LPP CCSD(T) value deviates only by 2.2° (1.9%). However, in the case of LuF$_3$ the AE DFT bond angle ($\angle$=101.4°) [181] is by more than 18° smaller than the LPP CCSD(T) result, i.e. at the AE DFT level a non-planar LuF$_3$ structure is obtained, whose bond angle is even smaller than that of LaF$_3$. Clavaguera et al. [181] confirmed this small bond angle by a SPP MP2 calculation using (14s13p10d8f6g)/[10s8p5d4f3g] [183] and cc-pVDZ basis sets for Lu and F, respectively. This finding stands in contrast to all calculations performed here for LuF$_3$, i.e. LPP, SPP, and AE/DKH CCSD(T) optimizations [52] using C_{3v} symmetry yield planar structures (cf. Table 3.28). Furthermore, a planar AE/DKH CCSD structure (R_e=1.966 Å) was confirmed as a true energy minimum by a numerical vibrational frequency analysis [52]. In order to exclude that this discrepancy is due to different methods or basis sets, a SPP MP2 optimization in GAUSSIAN03 [184] using exactly the same basis sets as Clavaguera et al. [181] was performed as well. A slightly non-planar structure confirmed as a true energy minimum by a numerical vibrational frequency analysis was obtained, but the bond angle amounts to 119.0 and not to 101.4°. Therefore the recently published results from Roos et al. [185], where the LuF$_3$ molecule is found to be planar (AE/DKH results: CASPT2 R_e=1.961; DFT/B3LYP: R_e=1.985 Å) and the 4f shell is affirmed to be essentially inert, are confirmed.

From the comparison of the LPP CCSD(T) structures to the experimental data [175] one can conclude that the new basis sets yield reasonable results especially for bond

3.2 4F-IN-CORE PSEUDOPOTENTIALS FOR LANTHANIDES

lengths, where the deviations are at most 0.014 Å (0.7%). Moreover, the LPP HF and CCSD(T) results are as good as those from former LPP HF and MP2 calculations [178], respectively, where the original (7s6p5d)/[5s4p3d] basis sets [17] were applied. In contrast to AnF$_3$ (cf. Sect. 3.1.5.6), the CPPs show only at the CCSD(T) level a slight improvement of the structures.

Atomization and Ionic Binding Energy The LPP HF and SPP state-averaged MCSCF atomization energies for LnF$_3$ $\Delta E_{at} = E(Ln) + 3 \times E(F) - E(LnF_3)$ calculated with respect to the lanthanide atom in the lowest state of the 4f^{n}5d^{1}6s^2 configuration show rather large differences, i.e. the mean and maximum error amount to 0.38 (2.5%) and 0.85 eV (5.7%), respectively (cf. Fig. 3.15 and Table C.12). This is due to the different kinds of coupling between the 4f and 5d shell. In the case of the SPP calculations the 4f shell is treated explicitly and therefore the 5d shell is coupled in such a way that the lowest LS-state according to Hund's rule arising from 4f^{n}5d^{1}6s^2 is obtained. In the case of the LPP calculations only ^{2}D valence substates are calculated, since the 4f shell is included in the PP core. Thus, 4f intrashell and 4f–5d intershell coupling is treated in an averaged manner and the results are obtained for averages over all states belonging to the molecular Ln^{3+} 4f^n (F$^-$)$_3$ and atomic Ln 4f^{n}5d^{1}6s^2 superconfigurations.

In order to correct for this discrepancy, the AE WB energy differences between the energies, where 4f^n is in its lowest LS-state and the 5d electron is coupled in an averaged manner, and the energies of the lowest LS-states according to Hund's rule were determined [55]. If this energy differences are subtracted from the LPP atomization energies, LPP HF atomization energies with respect to the lowest LS-states according to Hund's rule $\Delta E_{coupl.}$ are obtained (cf. Table C.12). By this correction the mean and maximum deviations between LPP and SPP atomization energies are clearly reduced to 0.16 (1.0%) and 0.58 eV (3.6%), respectively (cf. Fig. 3.16).

For Ce (4f^{1}5d^1) and Yb (4f^{13}5d^1) this energy correction could also be determined more exactly by subtracting the SPP state-averaged MCSCF atomization energy calculated with respect to the lowest LS-state according to Hund's rule (Ce: ^{1}G 15.18, Yb: ^{3}H 16.16 eV) from the atomization energy calculated with respect to the averaged energy of all possible LS-states for 4f^{n}5d^1 (n=1, 13 for Ce, Yb), namely 35 triplet and 35 singlet microstates (^{3}P, ^{3}D, ^{3}F, ^{3}G, ^{3}H, ^{1}P, ^{1}D, ^{1}F, ^{1}G, ^{1}H; Ce: 15.88, Yb: 16.41 eV). In the first case the lowest SPP state-averaged MCSCF energy of CeF$_3$ and YbF$_3$ was

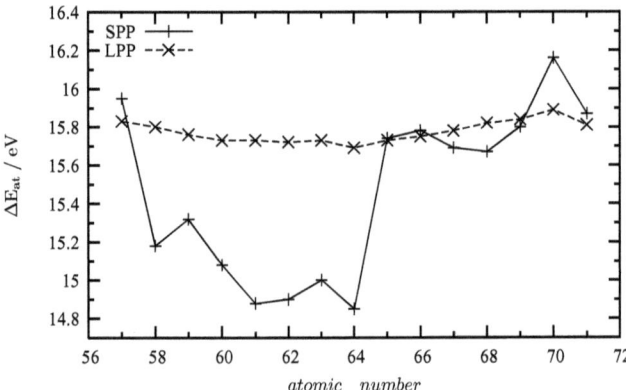

Figure 3.15: Atomization energies ΔE_{at} (in eV) for LnF$_3$ (Ln=La–Lu) with respect to the valence substates $4f^n 5d^1 6s^2$ from LPP HF and SPP state-averaged MCSCF calculations. SPP values are given for the lowest LS-states according to Hund's rule, while LPP values are given for ^2D states, since here the 4f shell is treated in an averaged manner.

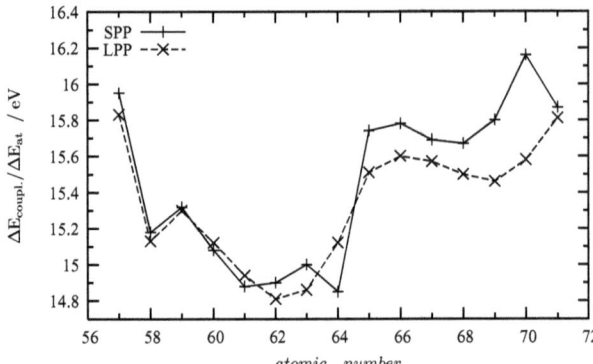

Figure 3.16: Atomization energies $\Delta E_{coupl.}$ and ΔE_{at} (in eV) for LnF$_3$ (Ln=La–Lu) with respect to the lowest LS-states according to Hund's rule corresponding to $4f^n 5d^1 6s^2$ from LPP HF and SPP state-averaged MCSCF calculations, respectively. The LPP atomization energies are corrected to account for the coupling between $4f^n$ and $5d^1$.

used to calculate the atomization energy, while for the latter case the average of these energies was taken. Here, the seven molecular energies of CeF$_3$ and YbF$_3$ corresponding to the ^{2}F state of 4f^1 and 4f^{13}, respectively, are not degenerate due to the ligand field generated by the fluorine atoms. The energy corrections for Ce and Yb are 0.70 and 0.25 eV, respectively, which is in good agreement with the AE WB corrections of 0.68 and 0.31 eV. Therefore the AE WB energy values should be a good approximation for these corrections.

Table 3.29: Atomization energies ΔE_{at} (in eV) for LnF$_3$ (Ln=La–Lu) from CCSD(T) calculations using LPPs with and without CPPs in comparison to experimental data [176] as well as to LPP CISD+Q [186] calculations. Additionally, LPP and LPP+CPP CCSD(T) atomization energies corrected to account for the coupling between 4f^n and 5d^1 as well as for the proper description of triply-charged ions Ln^{3+} $\Delta E_{coupl.+Ln^{3+}}$ (in eV) are given.

Ln	ΔE^a_{at}				$\Delta E^a_{coupl.+Ln^{3+}}$	
	LPP	CPP[b]	Exp.[c]	[186]	LPP	CPP[b]
La[d]	20.13	20.16	19.86	18.17	20.07	20.11
Ce	20.10	20.15	20.08	18.23	19.32	19.36
Pr	19.50	19.55	19.08	17.63	18.92	18.97
Nd	19.17	19.23	19.04	17.33	18.44	18.50
Pm	19.99	20.06			19.07	19.14
Sm	17.71	17.78	17.30	15.93	16.65	16.73
Eu	16.51	16.59	17.22	14.86	15.52	15.60
Gd	19.92	20.00	19.21	18.16	19.24	19.33
Tb	19.90	19.99	18.99	18.18	19.58	19.67
Dy	19.01	19.11	17.35	17.25	18.77	18.87
Ho	18.94	19.05	17.22	17.15	18.66	18.76
Er	19.14	19.25	17.17	17.31	18.76	18.87
Tm	18.43	18.53	17.04	16.55	18.00	18.11
Yb	17.26	17.36	16.05	15.29	16.91	17.02
Lu[d]	20.07	20.14	18.43	18.21	20.05	20.12

[a] Except for Pm, atomization energies are given with respect to the real ground states using experimental energy corrections (cf. Table C.11).
[b] LPP calculations using CPPs.
[c] The experimental data [176] were not measured directly, but calculated from a thermochemical cycle, where sometimes estimated values were used. The errors are assumed to be ±0.22 eV for La, Pr, Nd, Gd, and Er and ±0.43 eV for the other lanthanides. Further experimental values for ΔE_{at} (in eV): Sm 17.74(9), Eu 16.48(9), Tm 17.52(9) [177].
[d] SPP CCSD(T): La ΔE_{at}=20.38 eV; Lu ΔE_{at}=20.19 eV; AE/DKH CCSD(T): Lu ΔE_{at}=20.75 eV.

Since both LPPs and SPPs are only adjusted to reference configurations of neutral atoms and singly-charged cations [14, 17], the triply-charged cations may not be described accurately. To estimate this shortcoming, first the AE WB, LPP HF, and SPP HF energy differences between the configurations $5s^2 5p^6 5d^1 6s^2$ and $5s^2 5p^6$ of the neutral lanthanides and the triply-charged cations, respectively, were calculated [55]. Then the AE WB energies were subtracted from the LPP and SPP HF energies, respectively, to obtain the corresponding energy corrections. Although the obtained LPP and SPP atomization energies $\Delta E_{coupl.+Ln^{3+}}$ and $\Delta E_{at+Ln^{3+}}$ show a slightly larger m.a.e. and m.r.e. of 0.24 eV and 1.6%, respectively, the maximum deviation is reduced to 0.36 eV and 2.3% (cf. Fig. 3.17).

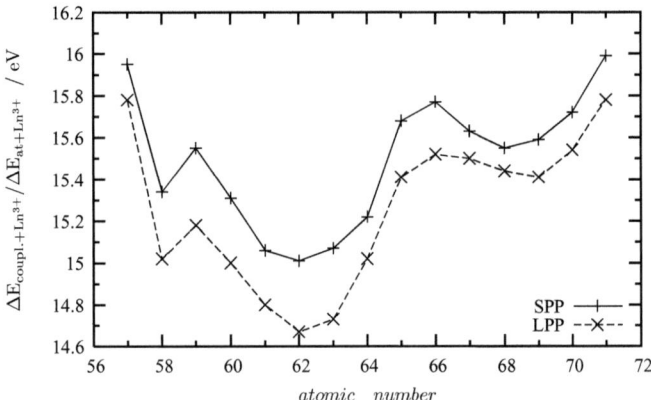

Figure 3.17: Atomization energies corrected to account for the proper description of triply-charged ions Ln^{3+} $\Delta E_{coupl.+Ln^{3+}}$ and $\Delta E_{at+Ln^{3+}}$ (in eV) for LnF_3 (Ln=La–Lu) with respect to the lowest LS-states according to Hund's rule corresponding to $4f^n 5d^1 6s^2$ from LPP HF and SPP state-averaged MCSCF calculations, respectively.

In order to directly compare LPP and SPP energies, ionic binding energies defined as $\Delta E_{ion} = E(Ln^{3+}) + 3 \times E(F^-) - E(LnF_3)$ were calculated. As one can see from Fig. 3.18 the LPP HF ionic binding energies underestimate the SPP values only by about 0.14 eV (0.3%). The effect of the CPPs grows with increasing nuclear charge, i.e. the ionic binding energies are increased by 0.12 and 0.86 eV for LaF_3 and LuF_3, respectively. Therefore the CPPs reduce the deviations from the SPP data for LaF_3–SmF_3 and increase them for EuF_3–LuF_3. Altogether, the mean deviation from the SPP

3.2 4F-IN-CORE PSEUDOPOTENTIALS FOR LANTHANIDES

results is by ca. 0.2 eV larger than that using pure LPPs, i.e. the m.a.e. (m.r.e.) amounts to 0.33 (0.7%). This is most likely due to the fact that CPPs include static and dynamic core-polarization effects, while the SPP state-averaged MCSCF calculations only account for the former.

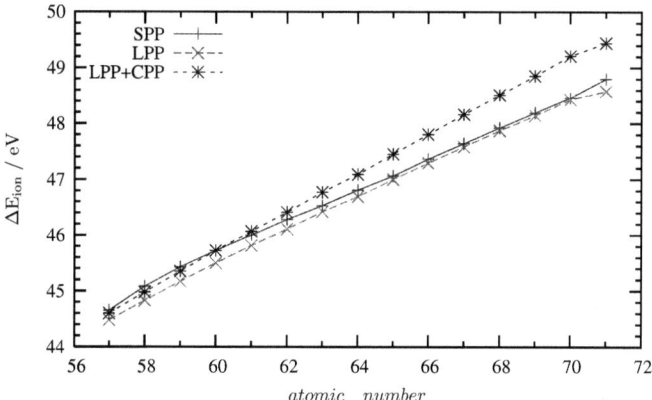

Figure 3.18: Ionic binding energies ΔE_{ion} (in eV) for LnF$_3$ (Ln=La–Lu) from LPP HF calculations with and without CPPs as well as from SPP state-averaged MCSCF calculations.

Since for LnF$_3$ experimental atomization energies [176, 177] are available, for the LPP CCSD(T) calculations atomization energies were determined. In order to compare the LPP CCSD(T) atomization energies to the experimental values, the energies have to be calculated with respect to the experimentally observed ground states. Therefore for those cases, where the $4f^n 5d^1 6s^2$ subconfiguration corresponds to an excited state, the atomization energies were corrected to account for the experimentally observed ground state by subtracting the energy separation taken from experiment [71] (cf. Table C.11). This energy difference could also be determined, e.g., at the AE WB [55] or AE DHF [44] level, whereby electron correlation contributions are neglected (cf. Tables C.9 and C.10 for AE WB and DHF corrections, respectively).

At the correlated level, the LPP CCSD(T) atomization energies ΔE_{at} deviate on average by 0.94 (5.4%) and 0.33 eV (1.9%) from the experimental atomization energies determined in 1975 [176] and 1981 [177], respectively. The application of CPPs has a very small effect on the atomization energies, i.e. they are increased by at most

0.11 eV (0.5%). Since in the case of the LPP CCSD(T) calculations the atomization energies are overestimated, the use of CPPs results in slightly increased deviations, i.e. the m.a.e. (m.r.e.) amount to 1.01 (5.7%) and 0.39 eV (2.2%) for the experimental atomization energies from 1975 and 1981, respectively. Former LPP calculations using (7s6p5d1f)/[5s4p3d1f] basis sets for Ln at the CISD+Q level [186] show larger differences to the experimental data than the LPP CCSD(T) calculations, i.e. the mean deviations are 1.00 (5.4%) and 1.47 eV (8.5%) for the experimental atomization energies from 1975 and 1981, respectively. The LPP CCSD(T) atomization energies are on average by 1.83 eV (10.7%) higher than those of the LPP CISD+Q calculations. However, this deviation is not necessarily due to the different basis sets for Ln, but probably results from the fact that in the older work [186] the original f-projector [17] not allowing any 4f participation in bonding was applied. Additional smaller deviations may result from the different core definitions and basis sets for F or the different molecular symmetries. Moreover, the size-extensivity of the CCSD(T) approach explains these results.

Since the experimental data are overestimated by up to 1.97 eV (11.5%), additional SPP CCSD(T) calculations for LaF$_3$ as well as LuF$_3$ and an AE/DKH CCSD(T) calculation for LuF$_3$ were carried out [52] (cf. Table 3.29). The obtained SPP and AE atomization energies are larger than the corresponding LPP and LPP+CPP results and thus overestimate the experimental values even more. The deviations between LPP and SPP atomization energies are very small, i.e. they amount to 0.25 (1.2%) and 0.12 eV (0.6%) for LaF$_3$ and LuF$_3$, respectively. Compared to the SPP data CPPs show an improvement of the pure LPP results, i.e. if CPPs are used, the differences to the SPP data are reduced to 0.22 (1.1%) and 0.05 eV (0.2%) for LaF$_3$ and LuF$_3$, respectively. Compared to the AE/DKH CCSD(T) atomization energy of LuF$_3$ the LPP and LPP+CPP CCSD(T) results are clearly too small, i.e. the differences amount to 0.68 (3.3%) and 0.61 eV (2.9%), respectively. These large deviations are most likely due to the BSSE, which tends to become larger with increasing number of explicitly treated electrons on the metal (LPP < SPP < AE).

Taking the BSSE into account via the counterpoise (CP) correction the LaF$_3$ atomization energies are reduced from 20.38 to 20.20 and from 20.13 to 19.98 eV for SPP and LPP calculations, respectively. In the case of LuF$_3$ these energies are reduced from 20.75 to 19.79, from 20.19 to 19.82, and from 20.07 to 19.81 eV for AE, SPP, and LPP calculations, respectively. As expected the BSSE using LPPs or SPPs are clearly smaller than that of the AE calculations, which constitutes an advantage compared to this more rigorous method. For the CP corrected atomization energies of LaF_3/LuF_3 the deviations between LPP and SPP values are reduced to 0.22/0.01 eV (1.1/0.05%), and the deviation between the LPP and AE value for LuF_3 is reduced to 0.02 eV (0.1%). Compared to the experimental data of LaF_3/LuF_3 from 1975 [176] the differences of the CP corrected energies amount to —/+1.36, +0.34/+1.39, and +0.12/+1.38 eV for AE, SPP, and LPP calculations, respectively.

Two possible reasons, why the experimental values are overestimated, are the neglect of SO effects and of the zero-point energy. The SO splittings for La and Lu can be taken from experiment [71] and amount to ca. 0.08 and 0.15 eV, respectively. The zero-point energy is estimated to be 0.13 eV, which corresponds to the value determined by LPP MP2 calculations for ErF_3 and TmF_3 [187]. If these corrections are added to the experimental data [176], the atomization energies for LaF_3 and LuF_3 are 20.07±0.22 and 18.71±0.43 eV, respectively. Compared to these energies for LaF_3/LuF_3 the deviations of the CP corrected values amount to —/+1.08, +0.13/+1.11, and −0.09/+1.10 eV for AE, SPP, and LPP calculations, respectively. While the computational atomization energies for LaF_3 are within the experimental error bars, those for LuF_3 are still by up to 1.11 eV (5.9%) too large. Since the more rigorous SPP and AE methods are known to be reliable, the uncertainty of the experimental data, for which estimated values had to be used within the thermochemical cycle, are most likely larger than the assumed 0.43 eV.

In the case of the other lanthanide trifluorides CeF_3–YbF_3 aside these reasons for the large discrepancies, the LPP atomization energies are also somewhat too large due to the wrong coupling between $4f^n$ and $5d^1$ as well as to the bad description of triply-charged ions (cf. above). If this is taken into account using the energy corrections determined at the HF level ($\Delta E_{coupl.+Ln^{3+}}$ in Table 3.29), the mean deviations compared to the experimental values from 1975 [176] are slightly reduced from 0.94 (5.4%) to 0.90 eV (5.1%). However, compared to the experiment from 1981 [177] the mean differences are increased by more than 50% from 0.33 (1.9%) to 0.84 eV (4.9%). Two possible reasons for the remaining deviations are the calculation of the energy corrections at the HF instead of the CCSD(T) level and the correction with respect to the lowest LS-states according to Hund's rule, which do not always correspond to the lowest experimentally observed LS-states. The AE WB corrections were calculated for LS-states according to Hund's rule, since some LS-states, e.g., ^4I of Pr $4f^25d^1$, cannot be calculated using the program MCHF95 [55], because there is more than one possibility to couple, e.g., $4f^2$ and $5d^1$ to obtain ^4I.

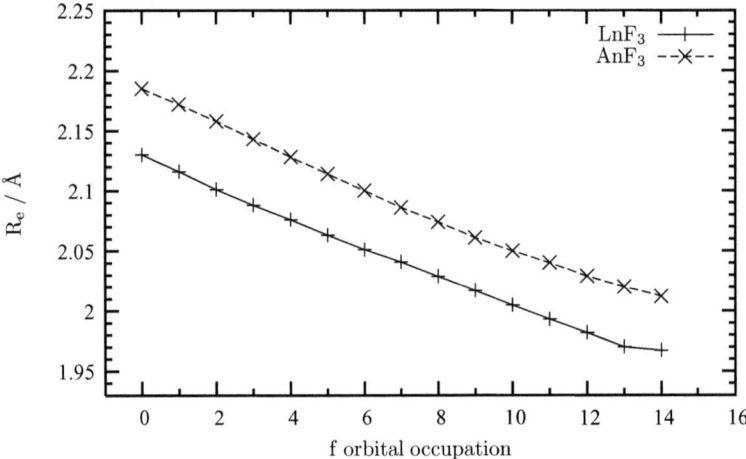

Figure 3.19: Comparison between M–F (M=Ln, An) bond lengths R_e (in Å) for LnF$_3$ (Ln=La–Lu) and AnF$_3$ (An=Ac–Lr) from LPP+CPP CCSD(T) calculations. La–Lu and Ac–Lr correspond to 0–14 f electrons.

Comparison between LnF$_3$ and AnF$_3$ In Fig. 3.19 the LPP+CPP CCSD(T) bond lengths for LnF$_3$ (Ln=La–Lu) and AnF$_3$ (An=Ac–Lr) are shown. One can see that both the Ln–F and An–F bond lengths decrease nearly linearly with increasing nuclear charge by 0.16 and 0.17 Å, respectively, which is due to the lanthanide/actinide contraction. Furthermore, the Ln–F and An–F bond lengths are almost parallel, whereby the An–F distances are on average 0.05 Å longer because of the bigger atomic radii. In the case of LnF$_3$, the LPP bond lengths are in good agreement with experiment, where all lanthanides are found to be trivalent. Since the Ln–F and An–F bond lengths are parallel to each other, the LPP results for ThF$_3$ and PaF$_3$ should also be reasonable, if for these compounds a trivalent oxidation state would be present. Thus, the 5f-in-core approximation only fails for ThF$_3$ and PaF$_3$ due to the fact that here the assumption of a trivalent oxidation state is not realized. However, this does not mean that the trivalent LPPs for Th and Pa are inaccurate, e.g., in the case of the actinide(III) mono- [20] and polyhydrates [24], where Th and Pa are in fact trivalent, the LPP results for these

3.2 4F-IN-CORE PSEUDOPOTENTIALS FOR LANTHANIDES

elements are also reliable.

As one can see from Fig. 3.20 the LPP+CPP CCSD(T) bond angles for AnF$_3$ (An=Ac–Lr) are obviously smaller than those for LnF$_3$ (Ln=La–Lu) and amount on average to 112.0°. Thus, for all AnF$_3$ C_{3v} symmetry is found, while in the case of LnF$_3$ for the lighter (La–Dy; ten elements) and heavier lanthanides (Ho–Lu; five elements) C_{3v} and D_{3h} symmetry is obtained, respectively. The reason for the larger F–Ln–F bond angles is most likely the smaller lanthanide atomic radii, due to which the F atoms come closer together than for AnF$_3$ and try to avoid each other by increasing the bond angles. This could also explain why the bond angles become larger with increasing nuclear charge, because the higher the nuclear charge the smaller the atomic radii.

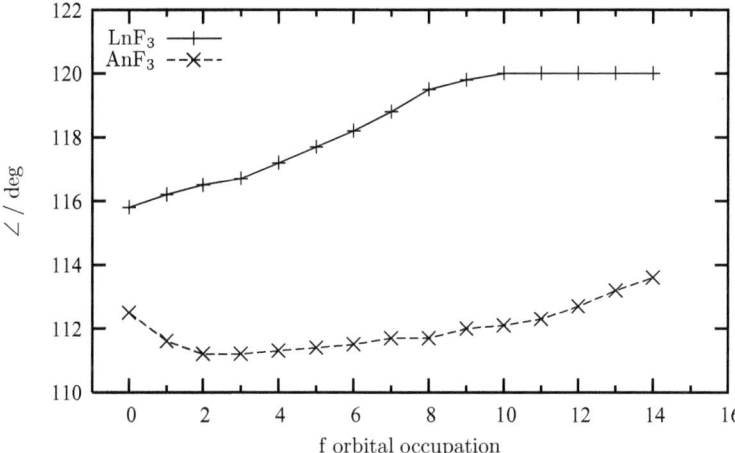

Figure 3.20: Comparison between F–M–F (M=Ln, An) bond angles ∠ (in deg) for LnF$_3$ (Ln=La–Lu) and AnF$_3$ (An=Ac–Lr) from LPP+CPP CCSD(T) calculations. La–Lu and Ac–Lr correspond to 0–14 f electrons.

The LPP+CPP CCSD(T) ionic binding energies for LnF$_3$ (Ln=La–Lu) and AnF$_3$ (An=Ac–Lr) are presented in Fig. 3.21. The ionic binding energies increase with increasing nuclear charge and thus with decreasing bond lengths, i.e. for LaF$_3$/LuF$_3$ and AcF$_3$/LrF$_3$ the ionic binding energies amount to 45.41/49.80 and 43.91/48.72 eV, respectively. Analogous to the bond lengths the ionic binding energies are almost parallel, whereby those of LnF$_3$ are on average by 1.24 eV larger than those of AnF$_3$. This is most likely due to the smaller Ln–F bond lengths according to the increasing energy with increasing nuclear charge, i.e. the smaller the metal–fluorine distance the higher the ionic binding energy.

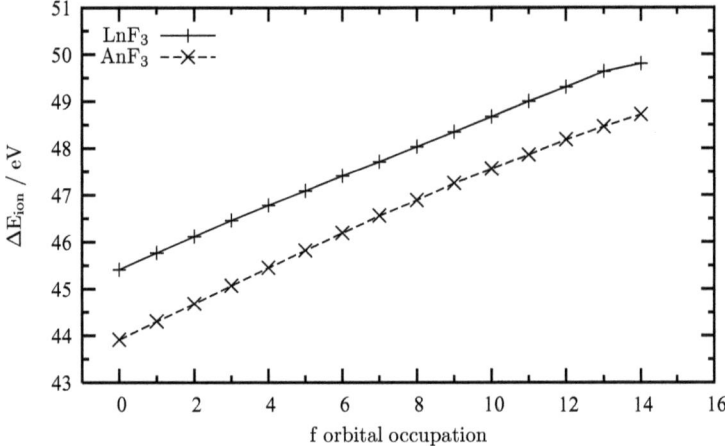

Figure 3.21: Comparison between ionic binding energies ΔE_{ion} (in eV) for LnF$_3$ (Ln=La–Lu) and AnF$_3$ (An=Ac–Lr) from LPP+CPP CCSD(T) calculations (LnF$_3$ values are given in Table C.13). La–Lu and Ac–Lr correspond to 0–14 f electrons.

3.3 5f-in-valence Pseudopotential for Uranium

In this Section test calculations for the recently adjusted MCDHF/DCB 5f-in-valence SPP for uranium and for the corresponding valence basis set will be described. First, the adjustment of the PP and the optimization of the basis set will be given [16]. Then test calculations for the U^{5+} SO splitting and the U^{4+} fine-structure spectrum [188] will be discussed.

3.3.1 Adjustment of the Pseudopotential

The new 5f-in-valence PP corresponding to uranium treats 32 valence electrons explicitly, while the 1s–4f shells (60 electrons) are included in the PP core. The PP parameters up to f symmetry were adjusted to four-component AE MCDHF/DCB[13] Fermi nucleus reference data [44], which comprised 100 non-relativistic configurations yielding a total of 30190 J levels. The reference data was obtained for U–U^{7+} and included a wide spectrum of occupations in the 5f, 6d, 7s, and 7p valence shells, but also additional configurations with holes in the core/semi-core orbitals 5s, 5p, 5d, 6s, and 6p as well as configurations with electrons in the 6f–9f, 7d–9d, 8p–9p, and 8s–9s shells (cf. Table A.11). Since the energetic position of the bare inner core relative to valence states is not expected to be notably relevant for chemical processes, the fit was restricted to the chemically more significant energy differences between valence states, i.e. a global shift was applied to all reference energies and treated as an additional parameter to be optimized [47] (cf. Sect. 2.1.6). The weights in (2.31) were chosen to be equal for all J levels arising from a non-relativistic configuration and all non-relativistic configurations were assigned to have equal weights. The g-part of the PP was adjusted to the eight energetically lowest U^{31+} [Kr]$4d^{10}4f^{14}ng^1$ (n=5–12) configurations. The PP parameters of the two-component form analogous to (2.25) are listed in Table A.12. The scalar-relativistic one-component as well as the SO potential may be derived using (2.28) and (2.29), respectively.

The adjustment of the s-, p-, d-, and f-parts shows mean square errors of 16.1 and 306.3 cm^{-1} for configurations (cf. Fig. 3.22) and J levels (cf. Fig. 3.23), respectively. The maximum absolute error amounts to 0.267 eV and is found for the 1218th J level of U [Rn]$5f^47s^17p^1$. The g-part can be considered as exact, i.e. the m.a.e. for the eight configurations amounts to 0.3 cm^{-1} corresponding to 0.00004 eV.

---

[13]The Breit interaction was included perturbatively.

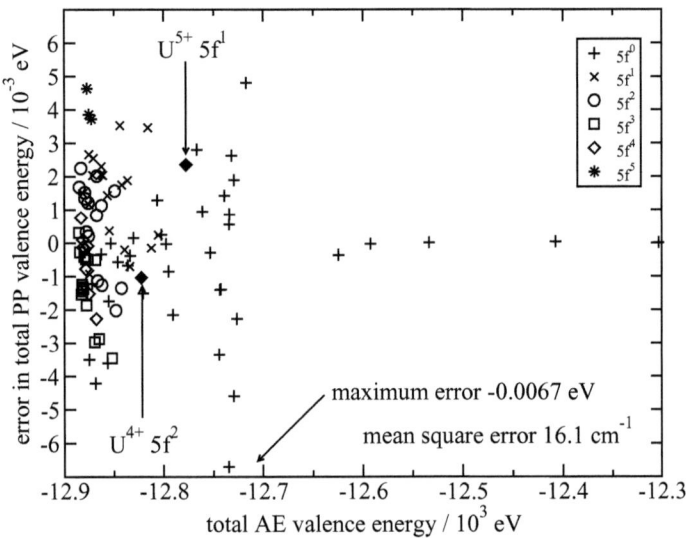

Figure 3.22: Errors in total valence energies (in eV) of 100 non-relativistic configurations for the MCDHF/DCB SPP for uranium [188]. The configurations U^{5+} $5f^1$ and U^{4+} $5f^2$ considered in the test calculations (cf. Sect. 3.3.3) are marked by filled diamonds.

3.3.2 Optimization of the Valence Basis set

The basis set optimization comprised four steps [16]. First, (14s11p8d8f) primitive Gaussians were HF energy-optimized [57] for the U [Rn]$5f^4 7s^2$ ^{5}I state. Secondly, two diffuse d and p functions were HF energy-optimized [57] for U [Rn]$5f^3 6d^1 7s^2$ ^{5}L and U [Rn]$5f^3 7s^2 7p^1$ ^{5}K, respectively. Thirdly, the atomic natural orbital (ANO) contraction coefficients for the resulting (14s13p10d8f)/[6s6p5d4f] set were obtained from averaged density matrices for the lowest LS-states of U [Rn]$5f^3 6d^1 7s^2$ and U [Rn]$5f^4 7s^2$ [52]. In the case of U [Rn]$5f^4 7s^2$ it was possible to perform a CASSCF calculation with a subsequent MRCI calculation (5s, 5p, and 5d shells were frozen), whereas U [Rn]$5f^3 6d^1 7s^2$ could only be treated at the CASSCF level. Finally, six g exponents were chosen identically to the six largest f exponents and a generalized ANO contrac-

3.3 5F-IN-VALENCE PSEUDOPOTENTIAL FOR URANIUM

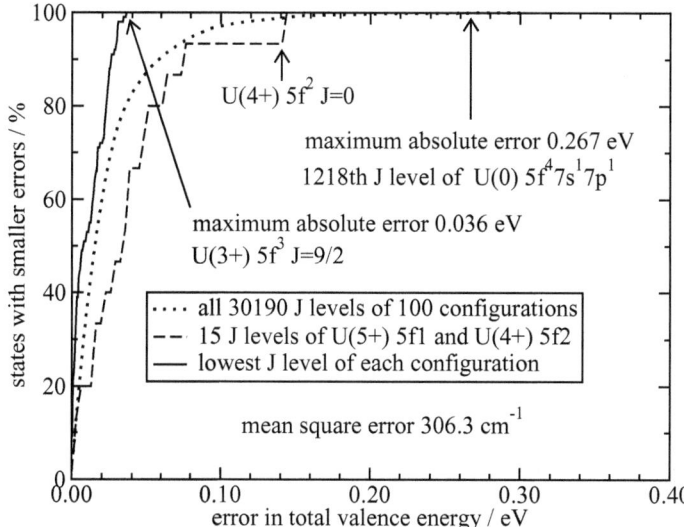

Figure 3.23: Percentage (%) of the J levels with errors in the total valence energy below the threshold (in eV) indicated on the abscissa for the MCDHF/DCB SPP for uranium [188]. The 15 J levels considered in the test calculations (cf. Sect. 3.3.3) are marked by a dashed line.

tion was derived for U [Rn]$5f^37s^2$ yielding the final (14s13p10d8f6g)/[6s6p5d4f3g] set of roughly pVQZ quality. The basis set parameters are compiled in Table B.14.

3.3.3 Atomic Test Calculations on U^{5+} and U^{4+}

In order to further test the MCDHF/DCB SPP for uranium, it was applied to calculate the SO splitting of U^{5+} ($5f^1$ subconfiguration) and the fine-structure spectrum of U^{4+} ($5f^2$ subconfiguration) [188] using a two-step CI approach identical to that used by Danilo et al. [189], i.e. a dressed effective Hamiltonian relativistic spin–orbit configuration interaction (SO-CI) scheme. These benchmark systems were chosen, because experimental [190–192] as well as computational [189, 193–196] reference data are available.

The study covers only 2% of the configurations and 0.04% of the J levels of the reference data used in the PP adjustment, and thus the applied PP is by no means tuned to describe these cases especially well. Figures 3.22 and 3.23 summarize the accuracy of the fit for the 100 configurations and 30190 J levels, respectively. The data for the U^{5+} $5f^1$ and U^{4+} $5f^2$ configurations considered here are marked specifically in these plots. It is obvious that the PP fit was not more accurate for these two cases than for the other configurations used in the reference data. Thus, the quality of the results obtained here should be representative of what could be obtained for other configurations, provided such large-scale calculations as presented here for U^{5+} $5f^1$ and U^{4+} $5f^2$ become generally feasible.

Electron correlation will be treated using the MRCI method with and without the Davidson size-extensivity correction (DaC) and with different frozen-orbital spaces. The best choice of the SO-CI configuration space is investigated in the case of the U^{5+} SO splitting. In addition results of intermediate Hamiltonian Fock-space coupled cluster (IH-FSCC) PP calculations using the approach of Kaldor and coworkers [197, 198] will be reported.

3.3.3.1 Computational Details

U^{5+} has only one 5f electron with a SO free ^2F ground state. Since at the SO free level of the calculation only one reference state is present, dynamic correlation cannot be included via an effective Hamiltonian dressing procedure. The orbital basis was obtained from a CASSCF calculation using the MCDHF/DCB PP and the corresponding (14s13p10d8f6g)/[6s6p5d4f3g] basis set on ^2F in MOLCAS [199] and the SO splitting between the $^2F_{5/2}$ ground and the $^2F_{7/2}$ excited state was calculated at the MCDHF/DCB PP SO-CI level using the EPCISO program [200]. In order to determine the best choice of the SO-CI configuration space, five different calculations were performed: diagonalizing the reference only (no single excitations, No S), adding single excitations from the 5f orbitals (S-f), adding single excitations from the 5d and 5f orbitals (S-df), adding single excitations from the 6p and 5f orbitals (S-pf), and including all single excitations from the 6p, 5d, and 5f orbitals (S-pdf). Note that for an atom single excitations from s shells do not contribute.

All possible LS-states of U^{4+} ($5f^2$ subconfiguration) were calculated at the state-averaged CASSCF/MRCI level [52] with and without DaC using the MCDHF/DCB PP and the corresponding (14s13p10d8f6g)/[6s6p5d4f3g] basis set (^1G, ^1D, ^1I, ^1S, ^3H, ^3F, ^3P). As comparison the same calculations were also performed using the old 5f-in-valence WB PP [15] and its corresponding (14s13p10d8f6g)/[6s6p5d4f3g] basis set [21]. Two different kinds of frozen-orbital spaces were used, i.e. while the 6s and 6p orbitals were always correlated, the 5s, 5p, and 5d orbitals were either frozen or correlated.

In the SO-CI calculation [200] of U^{4+} all 91 determinants generated by distributing two electrons in the 5f shell were included. The orbital basis was obtained from a MCDHF/DCB PP CASSCF calculation [199] on the triplet states of U^{4+}, where the orbitals were simultaneously optimized for all states. The (14s13p10d8f6g)/[6s6p5d4f3g] basis set corresponding to the PP was used and as configuration space the reference including all single excitations from the 6p, 5d, and 5f orbitals (S-pdf) was chosen, because it gave the best results for the U^{5+} SO splitting (cf. Sect. 3.3.3.2). The lacking dynamic correlation was included via a dressed effective Hamiltonian approach defined by the projection of the correlated SO free PP CASSCF/MRCI energies onto the SO-CI space [200, 201]. SO integrals were calculated with the semi-local relativistic effective SO operators corresponding to the MCDHF/DCB PP. As comparison analogous calculations were carried out using the old WB PP [15] and its corresponding effective SO operators designed for variational calculations [21] and the (14s13p10d8f6g)/[6s6p5d4f3g] basis set [21].

In addition to the SO-CI calculations IH-FSCC PP calculations using the approach of Kaldor and coworkers at the AE DHF/DCB level [193, 196–198] were performed. The basis set ranged from the (14s13p10d8f6g)/[6s6p5d4f3g] standard basis set to subsets of a (16s15p12d10f8g7h7i) uncontracted set obtained from the former primitive set by adding diffuse and higher angular momentum functions. The addition of further s to i functions to this set changes the fine-structure splittings of U^{5+} and U^{4+} by less than 1 cm^{-1}. All explicitly treated electrons were correlated and excitations in all virtual orbitals were allowed. The primary model space consisted of the 5f, 6d, and 7s orbitals, whereas the intermediate model space comprised the 6f–8f, 7d–9d, 8s–10s, 7p–10p, 5g–6g orbitals, i.e. all orbitals with negative energy for the standard contracted basis set in the U^{6+} reference system were included in the model spaces. Although larger intermediate spaces were not feasible due to the current hardware, almost identical results are obtained for an intermediate space reduced by one orbital for s to g symmetry, i.e. the U^{5+} and U^{4+} fine-structure splittings show mean average deviations of 1 and 10 cm^{-1}, respectively. Thus, the results presented here should not change significantly upon further increasing the intermediate space.

3.3.3.2 U^{5+} Spin–orbit Splitting

The SO splittings from SO-CI calculations with different definitions of the configuration space and IH-FSCC calculations with different basis sets using the MCDHF/DCB PP are listed in Table 3.30. In accordance with Hund's rules the ground state of U^{5+} is $^2F_{5/2}$. The experimental excitation energy for the $^2F_{5/2}$ to $^2F_{7/2}$ excitation or the experimental SO splitting amounts to 7609 cm^{-1} [190]. The up to now best theoretical value is 7598 cm^{-1} and deviates only by -11 cm^{-1} from the experiment. It was calculated by Infante et al. [193] using an AE DCB extrapolated IH-FSCC (XIH-FSCC) method [202].

AE/DKH SO-CI calculations by Danilo et al. [189] using the AMFI (atomic meanfield integral) code [203] implemented in MOLCAS [199] to determine the SO integrals are also available for comparison. Analogous to these SO-CI calculations five different configuration spaces were used, in order to find out the best choice, i.e. no single excitations (No S), single excitations from 5f (S-f), single excitations from 6p and 5f (S-pf), single excitations from 5d and 5f (S-df), and single excitations from 6p, 5d, and 5f (S-pdf). For all configuration spaces the PP results are in a range of 105–194 cm^{-1} smaller than the AE/DKH results. However, these deviations are not only due to the PP approach, but are also connected with the different basis sets (AE (26s23p17d13f5g)/[10s9p7d5f3g], PP (14s13p10d8f6g)/[6s6p5d4f3g]) and relativistic treatments (AE/DKH, PP DCB). Although both basis sets are of pVQZ quality, the AE set is based on CASPT2 ANOs and the PP set on CASSCF/MRCI ANOs. At the AE MCDHF level, using a Fermi nucleus, the results for the DC and DCB Hamiltonians are 7394 and 7083 cm^{-1}, respectively [44]. The energy difference of 311 cm^{-1} explains in part why also at the correlated level the PP results (modeling DCB) are lower than the DKH results (modeling DC). However, one also has to take into account that in the PP fit the total valence energy of $^2F_{5/2}$ is by 23 cm^{-1} too low, whereas the one of $^2F_{7/2}$ is by 50 cm^{-1} too high, resulting in an overestimation of the splitting by 73 cm^{-1} at this level of theory. The sum of these counteracting contributions suggests that the PP DCB splitting has to be by about 200 cm^{-1} lower than the AE/DKH value.

In accordance with the AE/DKH calculations [189] the smallest errors with respect to the experimental value occur using either no single excitations or including all single excitations from the doubly occupied orbitals in the near valence region, i.e. from 6p, 5d, and 5f (error: No S 219, S-pdf 183 cm^{-1} (2.4%)). Due to the implicit inclusion of the Breit interaction the PP results deviate even slightly less from the experimental

3.3 5F-IN-VALENCE PSEUDOPOTENTIAL FOR URANIUM

Table 3.30: SO splittings $\Delta E(SO)$ (in cm^{-1}) of the ^2F ground state of U^{5+} from SO-CI calculations using the MCDHF/DCB PP and the AE/DKH+AMFI method [189] as well as from MCDHF/DCB PP IH-FSCC calculations in comparison to experimental data [190] and AE DCB XIH-FSCC [193] results. Δ(exp) are deviations of the theoretical values from the experimental result and Δ(AE/PP) are deviations between the AE DCB XIH-FSCC or AE/DKH and MCDHF/DCB PP results.

Method[a]	$\Delta E(SO)$	Δ(AE/PP)	Δ(exp)
Exp.	7609		
DCB XIH-FSCC	7598	13	−11
DCB PP IH-FSCC spdfghi	7611		2
DCB PP IH-FSCC std.	7609		0
DCB PP No S	7828	194	219
AE/DKH No S	8022		413
DCB PP S-f	7868	180	259
AE/DKH S-f	8048		439
DCB PP S-pf	8407	189	798
AE/DKH S-pf	8596		987
DCB PP S-df	7213	105	−396
AE/DKH S-df	7318		−291
DCB PP S-pdf	7792	114	183
AE/DKH S-pdf	7906		297

Basis sets: PP std.: (14s13p10d8f6g)/[6s6p5d4f3g], spdfghi: (16s15p12d10f8g7h7i); AE/DKH (26s23p17d13f5g)/[10s9p7d5f3g]; AE DCB XIH-FSCC (37s32p24d21f12g10h9i).
[a]Different definitions of the configuration space: diagonalization of the reference only (No S), adding single excitations from 5f (S-f), adding single excitations from 5d and 5f (S-df), adding single excitations from 6p and 5f (S-pf), and including all single excitations from 6p, 5d, and 5f (S-pdf).

value than the AE/DKH results (error: No S 413, S-pdf 297 cm^{-1} (3.9%)). On the basis of these findings the S-pdf configuration space was used to calculate the U^{4+} fine-structure spectrum.

The best PP results were obtained with the IH-FSCC approach of Kaldor and coworkers [197, 198], i.e. splittings of 7609 and 7611 cm^{-1} for the standard (14s13p10d8f6g)/ [6s6p5d4f3g] and a (16s15p12d10f8g7h7i) uncontracted basis set, respectively. These excellent results are certainly somehow fortuitous, since the mean square error of the PP fit to the AE MCDHF/DCB reference data is 306 cm^{-1} for 30190 J levels arising from 100 configurations of U to U^{7+}. On the other hand the number is quite stable with respect to changes of the basis set, i.e. one obtains 7575 and 7597 cm^{-1} for the uncontracted subsets of (16s15p12d10f8g7h7i) containing only up to g and h functions, respectively.

Finally, results obtained with the older WB PP [15] should be discussed. For this PP two valence SO-terms acting on 5f, 6d, 7p, and higher orbitals of these angular momenta were published [21], i.e. one for use in first-order perturbation theory, the other one for use in valence variational or large-scale valence SO-CI calculations. At the variational and perturbative finite-difference level [44] these operators lead for U^{5+} to SO splittings of 6379 and 6590 cm^{-1}, respectively. Using the variational operator the valence SO-CI (S-f) leads to a value of 6374 cm^{-1}, demonstrating that the orbital relaxation under the SO-term is very well recovered by the single excitations. However, compared to the experimental value of 7609 cm^{-1} these SO splittings are by about 15% too small. The reason for this are missing correlation contributions in the SO-CI (S-f) and probably frozen-core errors. Note that these SO-terms, as the WB PP itself, were adjusted to U and U^+ to reproduce MCDHF splittings arising from the 5f, 6d, and 7p shells and only configurations with a 5f occupation of two to four were considered. For U^{5+} a SO splitting of 7083 cm^{-1} is obtained at the AE MCDHF/DCB Fermi nucleus level, i.e. a value by 526 cm^{-1} smaller than the experimental value results from neglecting electron correlation effects. In contrast to this the new MCDHF/DCB PP tested here was adjusted also to U^{5+} and allows a variational or large-scale SO-CI treatment for all orbitals treated explicitly, thus explaining the better results.

3.3.3.3 U^{4+} Fine-structure Spectrum

In this Section first the SO free correlated energies using the new MCDHF/DCB PP for U^{4+} with 5f^2 valence subconfiguration will be discussed in comparison to WB PP and AE/DKH [189] calculations (cf. Tables 3.31 and 3.32). Next the SO effect will be taken into account and the results will be compared to WB PP calculations as well as to experimental [191, 192] and computational [189, 193–195] data from the literature (cf. Tables 3.33, 3.34, 3.35 and 3.36).

Spin–orbit Free Calculations At the SO free level the ground state of U^{4+} with 5f^2 valence subconfiguration is ^{3}H as expected by Hund's rules. The MCDHF/DCB PP MRCI method with and without size-extensivity correction (DaC) was used in connection with different frozen-orbital spaces, i.e. 5s, 5p, and 5d were either frozen or correlated. First, the effect of the DaC and then that of the frozen-orbital space each with respect to the AE/DKH calculations [189] will be discussed. Finally, the new

3.3 5F-IN-VALENCE PSEUDOPOTENTIAL FOR URANIUM

Table 3.31: Energy levels with respect to the lowest energy level ^{3}H (in cm^{-1}) of U^{4+} with 5f^2 valence subconfiguration computed at the SO free level using the MCDHF/DCB PP in MRCI calculations without DaC and two different frozen-orbital spaces, i.e. 5s, 5p, and 5d either (1) frozen or (2) correlated. As comparison AE/DKH MRCI results [189] and the mean absolute deviation (m.a.d.) between the AE/DKH and the PP data are given.

State	MRCI (1) DCB^a	MRCI (2) DCB^a	MRCI (2) AE^b
^{3}H	0	0	0
^{3}F	3319	3488	3441
^{1}G	4971	5407	5173
^{1}D	12296	12727	12260
^{3}P	16204	16480	16325
^{1}I	17550	17013	16713
^{1}S	38458	40320	39822
m.a.d.		283	

Basis sets: PP (14s13p10d8f6g)/[6s6p5d4f3g]; AE (26s23p17d13f5g)/[10s9p7d5f3g].
^{a}MCDHF/DCB PP calculations.
^{b}1s–5p orbitals were frozen.

MCDHF/DCB PP will be compared to the old WB PP.

The MRCI method yields always higher energy levels than the MRCI+DaC method. Especially the ^{1}S state is influenced by the size-extensivity correction, i.e. it is lowered by 2112 and 3080 cm^{-1} for 5s, 5p, and 5d frozen and correlated, respectively. For the other states the deviations are smaller than 1900 cm^{-1}. Compared to the AE/DKH data the calculation at the MRCI+DaC level shows better agreement, i.e. if 5s, 5p, and 5d are correlated, the m.a.e. for all energy levels amount to 283 and 179 cm^{-1} for MRCI and MRCI+DaC, respectively. Here, the definition of the m.a.e. from Danilo et al. [189] and Infante et al. [193] was used, i.e. the sum of the absolute deviations between the PP and AE/DKH energies was divided by the number, namely six, of excited levels. However, one should rather use the term mean absolute deviation (m.a.d.) instead of m.a.e., since in contrast to the new PP the AE/DKH calculations do not include the Breit interaction and therefore the differences do not only result from the PP approach.

The different frozen-orbital spaces influence the energy levels by at most 1862 cm^{-1}. At the MRCI level the ^{1}S state is shifted the most (1862 cm^{-1}), while at the MRCI+DaC

Table 3.32: Energy levels with respect to the lowest energy level ^{3}H (in cm^{-1}) of U^{4+} with 5f^2 valence subconfiguration computed at the SO free level using the MCDHF/DCB PP in MRCI calculations with DaC and two different frozen-orbital spaces, i.e. 5s, 5p, and 5d either (1) frozen or (2) correlated. As comparison WB PP MRCI+DaC as well as AE/DKH MRCI+DaC results [189] and mean absolute deviations (m.a.d.) between the AE/DKH and the PP data are given.

	MRCI+DaC					
	(1)			(2)		
State	DCB^a	WB^b	AE	DCB^a	WB^b	AE^c
^{3}H	0	0	0	0	0	0
^{3}F	3050	3149	2887	3078	3151	3040
^{1}G	4828	4970	4728	5227	5372	4994
^{1}D	11355	11693	10806	11312	11611	11260
^{3}P	14941	15341	14256	14588	14946	14495
^{1}I	16953	17440	16296	15959	16380	15762
^{1}S	36346	37430	35383	37240	38297	36780
m.a.d.	520	944		179	571	

Basis sets: PP (14s13p10d8f6g)/[6s6p5d4f3g]; AE (26s23p17d13f5g)/[10s9p7d5f3g].
^{a}MCDHF/DCB PP calculations.
^{b}WB PP calculations.
^{c}1s–5p orbitals were frozen.

level the ^{1}I as well as the ^{1}S states are displaced by similar amounts, but in different directions, i.e. the ^{1}I state is lowered by 994 cm^{-1} and the ^{1}S state is increased by 894 cm^{-1}, respectively. At the MRCI+DaC level the PP results with 5s, 5p, and 5d correlated agree much better with the AE/DKH results (m.a.d. 179 cm^{-1}) than with 5s, 5p, and 5d frozen (m.a.d. 520 cm^{-1}). However, while in the latter case the AE and PP correlation spaces are exactly the same, they differ by frozen 5s and 5p shells at the AE level in the former case. The PP results obtained with 5s and 5p frozen exhibit a m.a.d. of 514 cm^{-1} to the corresponding AE values. For the PP SO-CI calculations reported below only the term energies with either 5s, 5p, and 5d frozen or correlated were considered.

If the old WB PP is used, the energy levels are always higher than those using the new MCDHF/DCB PP, whereby the increase grows from the ^{3}F to ^{1}S state, where it reaches more than 1000 cm^{-1}. Since the new PP already overestimates the AE/DKH energy levels, the m.a.d. are clearly larger using the old WB instead of the new MCDHF/DCB PP, i.e. the m.a.d. at the MRCI+DaC level using the new/old PP amount to 520/944

3.3 5F-IN-VALENCE PSEUDOPOTENTIAL FOR URANIUM

Table 3.33: Energy levels with respect to the lowest energy level $^3H_4$ (in cm^{-1}) and m.a.e. with respect to experimental [191, 192] data (cf. Table 3.35) of U^{4+} with 5f^2 valence subconfiguration. The SO free correlated energies obtained at the MRCI level using the MCDHF/DCB PP with different frozen-orbital spaces, i.e. 5s, 5p, and 5d either (1) frozen or (2) correlated, were used to dress the SO-CI matrix. Changes in the ordering of states are listed in italics.

J	Weight of LS-state^a	(1)	(2)
4	85% ^{3}H+9% ^{1}G	0	0
2	83% ^{3}F+11% ^{1}D	4919	5066
5	95% ^{3}H	6150	6110
3	95% ^{3}F	9544	9664
4	47% ^{3}F+40% ^{1}G	9790	10011
6	91% ^{3}H	11685	11628
2	54% ^{1}D+31% ^{3}P+10% ^{3}F	*18710*	*19011*
4	47% ^{3}F+46% ^{1}G	*17029*	*17262*
0	89% ^{3}P	20104	20409
1	95% ^{3}P	22695	22916
6	91% ^{1}I	25265	24734
2	63% ^{3}P+30% ^{1}D	27250	27512
0	89% ^{1}S	47970	49613
	m.a.e.	1689	1929

Basis sets: PP (14s13p10d8f6g)/[6s6p5d4f3g]; AE (26s23p17d13f5g)/[10s9p7d5f3g].
^{a}Weights from the MCDHF/DCB PP MRCI+DaC calculation with no frozen orbitals larger than 8% are given.

and 179/571 cm^{-1} for 5s, 5p, and 5d frozen and correlated, respectively.

Spin–orbit Calculations   The Tables 3.33 and 3.34 show the results for U^{4+} obtained at the SO-CI level by dressing the effective Hamiltonian matrix with MRCI(+DaC) correlated energies for the LS-states calculated using both the new MCDHF/DCB and old WB PP and two different frozen-orbital spaces. For comparison results of corresponding AE/DKH MRCI+DaC dressed effective Hamiltonian SO-CI calculations by Danilo et al. [189] are listed. The values included here are not the best results from this work (cf. Table 3.36), but merely those for which the correlation space fits best to the listed PP calculations. The second column of the Tables 3.33 and 3.34 presents information on the main SO free states contributing to each SO-state. The weights were obtained from a MCDHF/DCB PP calculation with a MRCI+DaC (no frozen orbitals) dressing of the effective SO Hamiltonian. Additionally, m.a.e. with respect to the experimental data [191, 192] (cf. Table 3.35) are given, which were calculated analogous to Danilo et al. [189] and Infante et al. [193], i.e. the sum of the absolute deviations between the calculated and experimental energy levels was divided by the number, namely 12, of excited levels. Analogous to the SO free calculations, first the effect of the size-extensivity correction (DaC) and then the difference between the two frozen-orbital spaces will be discussed. Next the results using the new MCDHF/DCB PP will be compared to those using the old WB PP as well as to AE/DKH calculations [189]. Finally, a comparison with other theoretical works [193–195] will be given (cf. Tables 3.35 and 3.36).

As for the SO free calculations the MRCI method yields higher energy levels than the MRCI+DaC method except for the $^3H_5$ state, which is slightly increased by 15 and 18 cm^{-1} for 5s, 5p, and 5d frozen and correlated, respectively, if the DaC is applied. Since the experimental energy levels are already overestimated by the MRCI+DaC results, the neglect of the size-extensivity correction leads to clearly increased deviations, i.e. the m.a.e. amount to 1689/1043 and 1929/948 cm^{-1} for MRCI/MRCI+DaC calculations with 5s, 5p, and 5d frozen and correlated, respectively. Except for the $^3H_5$ state all energy levels come closer to experiment and the largest improvement by 1939 (5s5p5d frozen) and 2841 cm^{-1} (5s5p5d correlated) is observed for the $^1S_0$ state. Thus, the size-extensivity correction gives a clear improvement of the results. Therefore the following discussions will be restricted to the MRCI+DaC method.

At the MRCI+DaC level the correlation of the 5s, 5p, and 5d orbitals leads to an improvement of the results, i.e. the m.a.e. decreases from 1043 to 948 cm^{-1}. This is due to the fact that the energy levels are lowered, if no orbitals are frozen, except for the two ($^3F,^1G)_4$ states and the $^1S_0$ state, where the energy levels are increased by 141, 155, and 741 cm^{-1}, respectively. The largest improvement due to the increased correlation space is obtained for the $^1I_6$ state and amounts to 941 cm^{-1}.

Compared to the old WB PP the new MCDHF/DCB PP seems not to improve the results, i.e. the m.a.e. differ by at most 4 cm^{-1}. However, this is most likely due to an error cancelation at the WB PP level, because the relativistic approach is more accurate for the new PP. The WB PP energy levels are always lower than those of the MCDHF/DCB PP and the maximum deviations are found for the higher ($^3F,^1G)_4$ state and amount to 2083 and 2074 cm^{-1} for 5s, 5p, and 5d frozen and correlated, respectively. While the new PP always overestimates the experimental energy levels, they

3.3 5F-IN-VALENCE PSEUDOPOTENTIAL FOR URANIUM

Table 3.34: Energy levels with respect to the lowest energy level $^3H_4$ (in cm^{-1}) and m.a.e. with respect to experimental [191, 192] data (cf. Table 3.35) of U^{4+} with 5f^2 valence subconfiguration. The SO free correlated energies obtained at the MRCI+DaC level using both the MCDHF/DCB PP and the WB PP with different frozen-orbital spaces, i.e. 5s, 5p, and 5d either (1) frozen or (2) correlated, were used to dress the SO-CI matrix. As comparison AE/DKH+AMFI MRCI+DaC data [189] are given. Changes in the ordering of states are listed in italics.

		(1)			(2)		
J	Weight of LS-state[a]	DCB[b]	WB[c]	AE	DCB[b]	WB[c]	AE[d]
4	85% ^{3}H+9% ^{1}G	0	0	0	0	0	0
2	83% ^{3}F+11% ^{1}D	4605	4508	4404	4585	4470	4559
5	95% ^{3}H	6165	5087	6406	6128	5055	6379
3	95% ^{3}F	9305	8290	9371	9295	8261	9490
4	47% ^{3}F+40% ^{1}G	9624	8665	9729	9765	8798	9883
6	91% ^{3}H	11681	9797	12054	11609	9742	12005
2	54% ^{1}D+31% ^{3}P+10% ^{3}F	*17803*	*16983*	*17434*	*17640*	*16795*	*17740*
4	47% ^{3}F+46% ^{1}G	*16851*	*14768*	*17192*	*17006*	*14932*	*17358*
0	89% ^{3}P	18873	18657	18157	18573	18298	18431
1	95% ^{3}P	21513	20677	21009	21142	20263	21210
6	91% ^{1}I	24733	23749	24395	23792	22716	23882
2	63% ^{3}P+30% ^{1}D	26172	24392	25969	25899	24066	26234
0	89% ^{1}S	46031	45066	45367	46772	45820	46602
	m.a.e.	1043	1046	887	948	952	1078

Basis sets: PP (14s13p10d8f6g)/[6s6p5d4f3g]; AE (26s23p17d13f5g)/[10s9p7d5f3g].
[a] Only weights from the MCDHF/DCB PP MRCI+DaC calculation with no frozen orbitals larger than 8% are given.
[b] MCDHF/DCB PP calculation.
[c] WB PP calculation.
[d] 1s–5p orbitals were frozen.

are both over- and underestimated by the old PP. Thus, the new PP seems at least to be more systematic in its deviations.

Analogous to the MCDHF/DCB PP the AE/DKH calculations always overestimate the experimental energy levels. If the 5s, 5p, and 5d orbitals are frozen, the PP energy levels are mainly larger than those of the AE/DKH calculations, and thus the m.a.e. is by 156 cm^{-1} larger than that of the AE/DKH values. If the 5s, 5p, and 5d orbitals are correlated, the PP energy levels are mainly lower than those of the AE/DKH calculations, and thus the m.a.e. of 948 cm^{-1} (5.1%) is by 130 cm^{-1} lower than that of the AE/DKH values, i.e. m.a.e. 1078 cm^{-1} (6.3%). However, the better agreement with

Table 3.35: Comparison of the best MCDHF/DCB PP results to experimental [191, 192] data for the energy levels with respect to the lowest energy level $^3H_4$ of U^{4+} with $5f^2$ valence subconfiguration (in cm^{-1}). Additionally, m.a.e. with respect to experimental data and m.a.d. with respect to AE DCB XIH-FSCC data are given. Changes in the ordering of states are marked in italics.

J	SO-CI^a	SO-CI^b	IH-FSCC^c	IH-FSCC^d	Exp.
4	0	0	0	0	0
2	4585	4406	3959	4233	4161
5	6128	6162	5902	5890	6137
3	9295	9191	8612	8825	8984
4	9765	9583	9196	9264	9434
6	11609	11608	11178	11144	11514
2	*17640*	*17195*	15998	*16601*	16465
4	*17006*	*16807*	16181	*16221*	16656
0	18573	18532	17025	17960	17128
1	21142	21112	19529	20420	19819
6	23792	23065	22594	22441	22276
2	25899	25659	24042	24799	24653
0	46772	46583	43783	45329	43614
m.a.e.	948	755	318	420	0
m.a.d.	628	436	567	162	357

^{a}MRCI+DaC + SO-CI, no frozen orbitals, (14s13p10d8f6g)/[6s6p5d4f3g] basis set.
^{b}IH-FSCC + SO-CI, no frozen orbitals, (14s13p10d8f6g)/[6s6p5d4f3g] basis set, LS-state energies from PP IH-FSCC using a (16s15p12d10f8g7h7i) basis set.
^{c}IH-FSCC, no frozen orbitals, (14s10p11d9f3g)/[6s5p5d4f1g] basis set.
^{d}IH-FSCC, no frozen orbitals, (16s15p12d10f8g7h7i) basis set.

the experimental data for 5s, 5p, and 5d correlated is probably due to the fact that in the AE/DKH calculation the 5s and 5p orbitals were frozen (cf. above). For both PP and AE/DKH calculations the largest deviations occur for the $^1S_0$ and $^1I_6$ states, whereby for the smaller correlation space the $^1I_6$ and for the larger one the $^1S_0$ state shows the larger deviation (5s5p5d frozen: PP: $^1I_6$ 2457, $^1S_0$ 2417; AE/DKH: $^1I_6$ 2119, $^1S_0$ 1753; 5s5p5d correlated: PP: $^1S_0$ 3158, $^1I_6$ 1516; AE/DKH: $^1S_0$ 2988, $^1I_6$ 1606 cm^{-1}).

Tables 3.35 and 3.36 summarizes the best PP results and compares them to selected computational ab initio data from the literature [189, 193, 195]. As mentioned above the m.a.e. with respect to the experiment of both PPs are almost the same, i.e. they amount to 948 (5.1%) and 952 cm^{-1} (7.2%) for the new and old PP, respectively. Thus, only the best results for the new PP from Table 3.34 are repeated for convenience.

3.3 5F-IN-VALENCE PSEUDOPOTENTIAL FOR URANIUM

Table 3.36: Comparison of AE results [189, 193, 195] to experimental [191, 192] data for the energy levels with respect to the lowest energy level $^3H_4$ of U^{4+} with $5f^2$ valence subconfiguration (in cm^{-1}). Additionally, m.a.e. with respect to experimental data and m.a.d. with respect to AE DCB XIH-FSCC data are given. Changes in the ordering of states are marked in italics.

J	DKH^a SO-CI	DCB^b MCDF+CI	DCB^c XIH-FSCC	Exp.
4	0	0	0	0
2	4501	3844	4202	4161
5	6392	6012	6070	6137
3	9455	8624	8974	8984
4	9819	9278	9404	9434
6	12010	11116	11420	11514
2	*17531*	15816	16554	16465
4	*17289*	15853	16630	16656
0	18170	16199	17837	17128
1	20960	18942	20441	19819
6	23744	22131	22534	22276
2	25998	23379	24991	24653
0	46189	43847	45611	43614
m.a.e.	935	522	357	0
m.a.d.	616	802	0	357

^{a}DKH MRCI+DaC + AMFI SO-CI, 1s–5p frozen, 5f–6d active, (26s23p17d13f5g)/[10s9p7d5f3g] basis set [189].
^{b}DCB MCDF-CI+DaC, (1s1p2d3f3g2h1i) spinors to describe the virtual space [195].
^{c}DCB XIH-FSCC, (37s32p24d21f12g10h9i) basis set [193].

The best calculations available so far for the fine-structure of the U^{4+} spectrum are the AE four-component calculations using the DCB Hamiltonian at the multi-configuration Dirac–Fock configuration interaction with Davidson correction (MCDF-CI+DaC) level [195] as well as at the XIH-FSCC level [193]. Both calculations yield the experimental ordering of the energy levels and show m.a.e. clearly below 1000 cm^{-1}, i.e. the m.a.e. amount to 522 (3.6%) and 357 cm^{-1} (1.5%) for the MCDF-CI+DaC and XIH-FSCC calculations, respectively. Therefore these calculations, especially FSCC, are useful as benchmark data, because they indicate the highest currently accessible accuracy for ab initio approaches in a theoretical spectrum for U^{4+}.

Even if the discrepancies between the AE four-component values and the PP SO-CI results are quite large, one should keep in mind the limitations of the AE four-component methods, i.e. the current implementation of the MCDF-CI can only be applied to atoms and the FSCC calculations to systems with at most two holes and/or electrons outside the reference closed-shell system [189]. The scope of the PP as well as the dressed effective Hamiltonian SO-CI starting from states obtained by correlated calculations within the Russell–Saunders scheme is much larger and both achieve a rather good agreement with the experimental values at a relatively low computational cost despite the strong SO coupling.

In order to separate the errors originating from the PP valence-only model Hamiltonian from those due to deficiencies of the one- and many-particle basis sets, the atomic DHF and FSCC suite of programs by Kaldor, Ishikawa, and coworkers [197, 198] was applied as well. The IH-FSCC results for the MCDHF/DCB PP are in quite satisfactory agreement with experimental data. Using the standard (14s13p10d8f6g)/[6s6p5d4f3g] basis set a m.a.e. (m.r.e.) with respect to experimental data of only 318 cm^{-1} (2.5%) is obtained, whereas the m.a.d. from the AE DCB XIH-FSCC results of Infante et al. [193] is 567 cm^{-1}. The corresponding results for the (16s15p12d10f8g7h7i) basis set are 420 (2.3%) and 162 cm^{-1}, respectively. Using subsets of the uncontracted basis set containing up to g and h functions, m.a.d. values with respect to the AE results of Infante of 432 and 204 cm^{-1}, respectively, are obtained.

Electron correlation effects are especially large for the $^1S_0$ state, which is calculated to be 673, 1430, and 1715 cm^{-1} above the experimental value of 43614 cm^{-1} for subsets of the uncontracted basis set containing up to g, h, and i functions, respectively. Although the errors with respect to experiment increase upon improving the correlation treatment, the result for the full basis set deviates by only 282 cm^{-1} from the AE DCB XIH-FSCC value 45611 cm^{-1}. A basis set extrapolation with respect to $1/l^3$, l being the highest angular quantum number present in the basis set, yields a value 2172 cm^{-1} above the experimental value. If it is taken into account that in the PP fit the $^1S_0$ state already is too high by 1144 cm^{-1} (cf. Fig. 3.23), the accordingly corrected estimate would be a term energy of 44642 cm^{-1}, i.e. the overestimation arising from the FSCC treatment could amount to about 1000 cm^{-1}. On the other hand it is fair to note that the $^1S_0$ state was originally not observed by Wyart et al. and its term energy has been semiempirically estimated to be 45812 cm^{-1} [191]. A refined estimate by Goldschmidt placed it at 45154 cm^{-1} [204]. Finally, van Deurzen et al. applied in their model Hamiltonian a γ parameter taken from the fitting of the spectra of trivalent actinides in crystals and estimated a value of 43480 cm^{-1} [192]. On the basis of this value three lines in the emission spectrum were interpreted as transitions from $^3D_1$, $^3P_1$, and $^1P_1$ of $5f^16d^1$ to $^1S_0$ of $5f^2$ and, applying the term energies of $5f^16d^1$ by Wyart et al. [191], the term energy of $^1S_0$ has been determined to be $43614\,\mathrm{cm}^{-1}$ [192]. More rigorous ab initio calculations than the ones reported here and in the literature so far are needed to clarify the situation.

Despite the smooth convergence of the PP results towards the AE results the ordering of the $(^3F,^1G)_4$ and $(^1D,^3P)_2$ levels near $16500\,\mathrm{cm}^{-1}$ is not reproduced correctly at the PP level with uncontracted basis sets, in contrast to the AE results. A notable exception is the standard contracted basis set, where the correct ordering of the states and the smallest error for the $^1S_0$ state is obtained. At the AE finite-difference MCDHF/DCB Fermi nucleus level $(^1D,^3P)_2$ is $5263\,\mathrm{cm}^{-1}$ above $(^3F,^1G)_4$ [44], whereas experimentally it is $191\,\mathrm{cm}^{-1}$ below. Thus, the calculations have to recover accurately quite differential correlation contributions. Otherwise the contributions to the fine-structure are quite accurately folded in the effective one-electron SO-term, which yields stable results already for comparatively small basis sets.

Deviations of a few hundred wavenumbers from experimental and highly accurate AE data have to be expected due to the accuracy of the PP adjustment. Thus, it is clear that an even better agreement with experimental data than obtained in the rigorous AE calculations is to a certain extent fortuitous. The question why the two-step SO-CI calculations exhibit larger errors was investigated by applying energies from IH-FSCC calculations using the scalar-relativistic part of the MCDHF/DCB PP and the (16s15p12d10f8g7h7i) basis set as diagonal elements in the SO-CI matrix, which was built using the standard basis set. The SO-CI and FSCC term energies up to 15000, between 15000 and 30000, and above $30000\,\mathrm{cm}^{-1}$ agree with a m.a.d. of 420, 1167, $2800\,\mathrm{cm}^{-1}$, respectively, if the standard basis set is applied in the FSCC. Corresponding m.a.d. of 319, 655, and $1254\,\mathrm{cm}^{-1}$ are obtained with respect to FSCC results calculated with the (16s15p12d10f8g7h7i) basis set. The m.a.e. (m.r.e.) for all levels listed in Table 3.35 with respect to experimental data is reduced from 948 (5.1%) to $755\,\mathrm{cm}^{-1}$ (3.8%), i.e. about 40% of the remaining error comes from the energies of the LS-states, whereas about 60% might be caused by insufficient orbital relaxation under the SO-term.

Barandiarán and Seijo [194] also performed a two-step method, where they used a CASPT2 calculation applying a WB ab initio model potential (1s–5d in core) to dress the SO-CI matrix. Their best results agree even better with the experiment than the AE four-component methods, i.e. the m.a.e. amounts to only 180 cm^{-1} (1.4%). However the reason for this very good agreement is, at least in part, due to parameter fitting, i.e. a scaling factor of 0.9 was applied to the WB SO operator and the ^{3}P and ^{1}I states were shifted downward by 1000 cm^{-1}. It is noteworthy that without these corrections the m.a.e. is much larger (1284 cm^{-1} (8.3%)), whereas the ordering of the states is entirely correct with (^{1}D,^{3}P)$_2$ being about 327 cm^{-1} below (^{3}F,^{1}G)$_4$ (experiment 191 cm^{-1}).

Ionization Potentials The fifth and sixth IP of U are obtained as a byproduct of the IH-FSCC calculations. The corresponding results are listed in Table 3.37. The values obtained with the contracted standard basis set deviate by only 543 and -279 cm^{-1} from the best corresponding AE data published by Infante et al. [193]. Using the uncontracted basis set and monitoring the behavior with respect to the highest angular momentum quantum number in the basis set one finds a systematic convergence, although the deviations from the AE values tend to become somewhat larger. Extrapolating the PP results linearly for the largest three basis sets with respect to $1/l^3$, with l being the highest angular momentum quantum number present in the basis set, values of 383960 and 510195 cm^{-1} are obtained for IP$_5$ and IP$_6$, respectively. The correlation coefficients deviate from the ideal value of -1 only by 2×10^{-5} and 2×10^{-6}, respectively. The extrapolated PP values are by 0.75 and 0.40% higher than the AE results, which are most likely also not fully converged with respect to the basis set. It should be noted here that experimental reference values do not exist.

3.3 5F-IN-VALENCE PSEUDOPOTENTIAL FOR URANIUM

Table 3.37: Ionization potentials IP_5 and IP_6 (in cm^{-1}) from MCDHF/DCB PP IH-FSCC calculations in comparison to corresponding AE DHF/DCB XIH-FSCC [193] results.

Method	IP_5	ΔIP_5	IP_6	ΔIP_6
DCB XIH-FSCC[a]	381074	0	508183	0
DCB PP std.[b]	381617	543	507904	−279
DCB PP spdf[c]	367431	−13643	491960	−16223
DCB PP spdfg[c]	378377	−2697	504299	−3884
DCB PP spdfgh[c]	381118	44	507170	−1013
DCB PP spdfghi[c]	382297	1223	508451	268
DCB PP ext.[c]	383960		510195	

[a] AE DCB XIH-FSCC (37s32p24d21f12g10h9i).
[b] (14s13p10d8f6g)/[6s6p5d4f3g] standard PP basis set.
[c] spdf ... spdfghi denotes a subset of a (16s15p12d10f8g7h7i) basis set; ext. denotes extrapolated values based on results obtained with the spdfg, spdfgh, and spdfghi basis sets.

Chapter 4
Conclusion and Outlook

This thesis completes earlier adjusted quasirelativistic energy-consistent f-in-core PPs for the f-elements, i.e. the missing 5f-in-core actinide PPs for di-, tetra-, penta-, and hexavalent oxidation states as well as 4f-in-core lanthanide PPs for tetravalent oxidation states were adjusted. Furthermore, corresponding molecular basis sets of pVDZ to pVQZ quality were optimized, whereby smaller basis sets suitable for calculations in crystalline solids form subsets of these basis sets. In order to account for the neglected static and dynamic core-polarization, CPPs were adjusted for di-, tri-, and tetravalent 5f-in-core actinide PPs. Finally, the LPPs, CPPs, and basis sets were tested in atomic and molecular test calculations and used for selected applications as actinocenes or uranyl(VI) complexes.

Besides the quasirelativistic f-in-core LPPs, atomic test calculations for a more rigorous 5f-in-valence SPP adjusted to AE four-component MCDHF/DCB reference data were performed. This PP includes scalar-relativistic effects as well as SO coupling into an effective one-electron Hamiltonian improving the relativistic treatment of the old WB 5f-in-valence SPP.

In the following the results for the 5f-in-core, 4f-in-core, and 5f-in-valence PPs will be summarized separately.

5f-in-core Pseudopotentials for Actinides Quasirelativistic 5f-in-core LPPs and corresponding valence basis sets for crystal and molecular calculations were adjusted for di- ($5f^{n+1}$, $n=5$–13 for Pu–No)[1], tetra- ($5f^{n-1}$, $n=1$–9 for Th–Cf), penta- ($5f^{n-2}$, $n=2$–6 for Pa–Am), and hexavalent ($5f^{n-3}$, $n=3$–6 for U–Am) actinide atoms. Fur-

---

[1] The parameters of the divalent 5f-in-core PPs I have already adjusted during my diploma thesis [19].

thermore, CPPs for di-, tri-, and tetravalent PPs were optimized.

Atomic test calculations for the first and second IPs show good agreement to experimental and SPP reference data except for those ionizations, where a change in the 6d or 7p shell occurs. However, this can be explained by the fact that LPPs show ionizations between averaged LS-states, while experimental and SPP ionizations take place between high-spin states. The application of CPPs gave clearly improved results, i.e. for the LPP/LPP+CPP calculations the m.a.e. for IP_1 and IP_2 amount to 0.40/0.33 and 0.20/0.10 eV, respectively.

Bond lengths from LPP HF and CCSD(T) calculations[2] on AnF_n (n=2–6) show good agreement to SPP and experimental reference data, respectively, except for PuF_2, ThF_3, and PaF_3, i.e. the maximum deviation without these compounds is 0.042 Å (2.0%). The reason why the a priori assumption of di- and trivalent actinides fails for PuF_2, ThF_3, and PaF_3 is most likely that for these actinides the di- and trivalent oxidation states are not preferred (Th) or even not stable (Pa, Pu) in aqueous solution. In the case of the bond energies reasonable results are only obtained for AnF_2–AnF_5 and UF_6 (largest deviation 0.29 eV (5.1%)), while for NpF_6–AmF_6 the 5f-in-core approximation becomes too crude (smallest deviation 0.69 eV (12.5%)). Thus, hexavalent 5f-in-core PPs of Np–Am should only be used for preoptimizing purposes. The application of CPPs improved the agreement for both bond lengths and energies, whereby the CPP effect decreases from di- via tri- to tetravalent LPPs because of the decreasing dipole polarizabilities. Therefore no CPPs for penta- and hexavalent oxidation states were adjusted.

The applications of the 5f-in-core LPPs to actinocene complexes and the bare uranyl ion show reasonable agreement with experimental and SPP reference data, i.e. the f-part of the LPPs are found to be able to model quite well the 5f orbital contributions to the bonding. However, in the case of the other penta- AnO_2^+ (An=U–Am) and hexavalent AnO_2^{2+} (An=Np–Am) actinyl ions the 5f-in-core approximation fails, because the a priori assumption of penta- and hexavalent oxidation states is not fulfilled. The LPP calculations[3] of the uranyl(VI) complexes with aromatic acids are in good agreement with experimental and SPP data for molecular structures and relative stabilities, but the TD-DFT excitation spectra do not show the important CT excitations from ligand π MOs to U 5f atomic-orbital-like MOs. This is due to the fact that the f-part of the LPP is already exhausted by the additional electrons due to the ligand-to-metal dona-

---
[2] The LPP HF calculation for BkF_4 and SPP calculations for ThF_4–CfF_4 were performed by X. Cao.
[3] The calculations were performed by D. Weißmann.

tion.

Finally, one can conclude that the di-, tri-, tetra-, and pentavalent 5f-in-core LPPs as well as the hexavalent uranium LPP are useful tools for those actinide compounds, where the 5f orbitals do not significantly contribute in chemical bonding. Thus, avoiding the difficulties due to the large number of electrons, the significant relativistic effects, and the open shells, many calculations especially on large actinide complexes become feasible. However, one should always explicitly test, if the 5f occupation number is close to integral, e.g. in single-point HF calculations with an explicit treatment of the 5f shell.

4f-in-core Pseudopotentials for Lanthanides Quasirelativistic 4f-in-core LPPs and corresponding valence basis sets[4] for use in crystal and molecular calculations were generated for tetravalent ($4f^{n-1}$, n=1–3, 8, 9 for Ce–Nd, Tb, Dy) lanthanide atoms. Results of LPP HF and CCSD(T) test calculations[5] for LnF$_4$ show reasonable agreement with SPP and experimental reference data, respectively, i.e. the maximum deviation for bond lengths and energies amount to 0.016 Å (0.8%) and 0.26 eV (5.1%), respectively. Furthermore, LPP HF and CCSD(T) calculations for CeO$_2$ are in satisfactory agreement with SPP reference data, if 3f basis functions HF optimized for $4f^2 5s^2 5p^6 6s^2$ instead of 2f polarization functions CI optimized for $5s^2 5p^6 5d^2 6s^2$ are applied. Thus, the tetravalent 4f-in-core LPPs are reliable tools to investigate compounds including tetravalent lanthanides within a reasonable amount of computer time, if the 4f shell does not participate significantly in bonding.

For LnF$_3$ 4f-in-core LPP HF calculations using the recently published (8s7p6d3f2g)/ [6s5p5d3f2g] basis sets show good agreement with corresponding SPP reference data, i.e. the mean deviations amount to 0.005 Å (0.2%), 0.2° (0.2%), and 0.14 eV (0.3%) for bond lengths, bond angles, and ionic binding energies, respectively. Compared to experimental data only LPP CCSD(T) bond lengths show small deviations, i.e. the m.a.e. for bond lengths, bond angles, and atomization energies are 0.007 Å (0.3%), 10.3° (9.5%), and 0.94 eV (5.4%), respectively. If corrections for the neglected SO coupling and zero-point energy as well as the BSSE are taken into account, the deviation for the atomization energy of LaF$_3$ lies within the experimental error bar, while that for LuF$_3$ still amounts to 1.10 eV (5.9%), which is more than 50% larger than the experimental error bar of 0.43 eV. The reason for the large discrepancies from ex-

---

[4] The tetravalent 4f-in-core PPs as well as the corresponding basis sets were adjusted by M Hülsen
[5] The calculations for LnF$_4$ were performed by M. Hülsen.

perimental energies might be that the assumed uncertainty for atomization energies including estimated values in the thermochemical cycle is too small. In the case of bond angles the experimental data neglect the anharmonicity in angle bending vibration, which can seriously affect the angle parameters. Thus, the rather large differences from experimental data for bond angles and atomization energies are not necessarily due to the LPP calculations, which are consistent with SPP/AE[6] CCSD(T) data for LaF$_3$ and LuF$_3$, i.e. the maximum deviations (using the counterpoise correction) are 0.4° (0.3%) and 0.22 eV (1.1%), both occurring for LaF$_3$. Therefore the test calculations demonstrate that the recently published basis sets yield reasonable results and are at least as good as the original ones.

5f-in-valence Pseudopotential for Uranium For the recently adjusted MCDHF/DCB 5f-in-valence SPP for uranium atomic test calculations on the SO splitting of U^{5+} and the fine-structure spectrum of U^{4+} were performed using a dressed effective Hamiltonian SO-CI framework as well as the IH-FSCC. The results were compared to those of the older WB PP, supplemented by a valence SO-term, as well as to AE/DKH calculations.

The SO-CI calculation for the SO splitting of U^{5+} yields good agreement with the experiment, i.e. the deviation is 183 cm^{-1} corresponding to 2.4%. Analogous to the AE/DKH+AMFI calculations the best results are obtained for the SO-CI configuration space, where all single excitations from the doubly occupied orbitals in the near valence region are included, i.e. from 6p, 5d, and 5f. Therefore this configuration space was used to calculate the fine-structure of the U^{4+} spectrum.

For the U^{4+} spectrum the SO-CI using the MCDHF/DCB PP with a MRCI+DaC dressing (no frozen orbitals) of the effective Hamiltonian gave the best results, i.e. the m.a.e. with respect to the experimental data amounts to 948 cm^{-1} corresponding to 5.1%. All energy levels are systematically overestimated by the new PP and the maximum deviation found for the $^1S_0$ state is 3158 cm^{-1}. Although the old WB PP yields a similar m.a.e. of 952 cm^{-1}, it is less systematic, because the energy levels are both over- and underestimated. The results of the new PP are in reasonable agreement with the AE/DKH MRCI+DaC calculations.

An accuracy very similar to the one obtained in AE four-component MCDF-CI+DaC and XIH-FSCC calculations is obtained, if the new PP is applied in the IH-FSCC

---

[6]The AE calculation for LuF$_3$ was performed by M. Dolg.

framework, i.e. for the largest basis set a virtually exact splitting for U^{5+} and a m.a.e. of 420 cm^{-1} for U^{4+} is found. In view of applications in larger molecular systems it is very promising that due to the folding of the relativistic effects including SO coupling into an effective one-electron Hamiltonian in the PP approach, results of this quality can already be obtained using standard contracted basis sets of pVQZ quality.

The atomic test calculations show that the MCDHF/DCB SPP for uranium yields reliable results and therefore confirm earlier benchmark calculations on uranium monohydride. Thus, the adjustment of analogous SPPs for the other actinide elements is reasonable, in order to have such improved relativistic PPs at hand for all actinides and to supplement the already existing MCDHF/DCB PPs for the main group and d elements.

Appendix A

Pseudopotentials

Table A.1: Reference configurations for the adjustment of the s-, p-, and d-parts of the di- ($5f^{n+1}$, n=5–13 for Pu–No; $Q = 10$) and tetravalent ($5f^{n-1}$, n=1–9 for Th–Cf; $Q = 12$) 5f-in-core PPs for actinides as well as of the tetravalent ($4f^{n-1}$, n=1–3, 8, 9 for Ce–Nd, Tb, Dy; $Q = 12$) 4f-in-core PPs for lanthanides.

	Actinides		Lanthanides
No.	$Q = 10$	$Q = 12$	$Q = 12$
1	$(5f^{n+1})\,6s^26p^67s^2$	$(5f^{n-1})\,6s^26p^66d^27s^2$	$(4f^{n-1})\,5s^25p^65d^26s^2$
2	$(5f^{n+1})\,6s^26p^6$	$(5f^{n-1})\,6s^26p^6$	$(4f^{n-1})\,5s^25p^6$
3	$(5f^{n+1})\,6s^26p^66d^1$	$(5f^{n-1})\,6s^26p^66d^1$	$(4f^{n-1})\,5s^25p^65d^1$
4	$(5f^{n+1})\,6s^26p^67s^1$	$(5f^{n-1})\,6s^26p^67s^1$	$(4f^{n-1})\,5s^25p^66s^1$
5	$(5f^{n+1})\,6s^26p^67p^1$	$(5f^{n-1})\,6s^26p^67p^1$	$(4f^{n-1})\,5s^25p^66p^1$
6	$(5f^{n+1})\,6s^26p^66d^2$	$(5f^{n-1})\,6s^26p^66d^2$	$(4f^{n-1})\,5s^25p^65d^2$
7	$(5f^{n+1})\,6s^26p^66d^17s^1$	$(5f^{n-1})\,6s^26p^66d^17s^1$	$(4f^{n-1})\,5s^25p^65d^16s^1$
8	$(5f^{n+1})\,6s^26p^66d^17p^1$	$(5f^{n-1})\,6s^26p^66d^17p^1$	$(4f^{n-1})\,5s^25p^65d^16p^1$
9	$(5f^{n+1})\,6s^26p^67s^17p^1$	$(5f^{n-1})\,6s^26p^67s^17p^1$	$(4f^{n-1})\,5s^25p^66s^16p^1$
10		$(5f^{n-1})\,6s^26p^67s^2$	$(4f^{n-1})\,5s^25p^66s^2$
11		$(5f^{n-1})\,6s^26p^66d^27s^1$	$(4f^{n-1})\,5s^25p^65d^26s^1$
12		$(5f^{n-1})\,6s^26p^66d^17s^2$	$(4f^{n-1})\,5s^25p^65d^16s^2$
13		$(5f^{n-1})\,6s^26p^67s^27p^1$	$(4f^{n-1})\,5s^25p^66s^26p^1$
14		$(5f^{n-1})\,6s^26p^67s^17p^2$	$(4f^{n-1})\,5s^25p^66s^16p^2$
15		$(5f^{n-1})\,6s^26p^66d^17s^17p^1$	$(4f^{n-1})\,5s^25p^65d^16s^16p^1$
16		$(5f^{n-1})\,6s^26p^66d^37s^1$	$(4f^{n-1})\,5s^25p^65d^36s^1$
17		$(5f^{n-1})\,6s^26p^66d^27s^17p^1$	$(4f^{n-1})\,5s^25p^65d^26s^16p^1$
18		$(5f^{n-1})\,6s^26p^66d^17s^27p^1$	$(4f^{n-1})\,5s^25p^65d^16s^26p^1$

Table A.2: Reference configurations for the adjustment of the s-, p-, and d-parts of the penta- ($5f^{n-2}$, n=2–6 for Pa–Am; $Q = 13$) and hexavalent ($5f^{n-3}$, n=3–6 for U–Am; $Q = 14$) 5f-in-core PPs for actinides.

No.	$Q = 13$	$Q = 14$
1	$(5f^{n-2})\,6s^26p^66d^37s^2$	$(5f^{n-3})\,6s^26p^66d^47s^2$
2	$(5f^{n-2})\,6s^26p^66d^2$	$(5f^{n-3})\,6s^26p^66d^3$
3	$(5f^{n-2})\,6s^26p^66d^17s^1$	$(5f^{n-3})\,6s^26p^66d^27s^1$
4	$(5f^{n-2})\,6s^26p^67s^2$	$(5f^{n-3})\,6s^26p^66d^17s^2$
5	$(5f^{n-2})\,6s^26p^66d^3$	$(5f^{n-3})\,6s^26p^66d^27p^1$
6	$(5f^{n-2})\,6s^26p^66d^27s^1$	$(5f^{n-3})\,6s^26p^66d^4$
7	$(5f^{n-2})\,6s^26p^66d^17s^2$	$(5f^{n-3})\,6s^26p^66d^37s^1$
8	$(5f^{n-2})\,6s^26p^66d^27p^1$	$(5f^{n-3})\,6s^26p^66d^27s^2$
9	$(5f^{n-2})\,6s^26p^66d^17s^17p^1$	$(5f^{n-3})\,6s^26p^66d^37p^1$
10	$(5f^{n-2})\,6s^26p^66d^37s^1$	$(5f^{n-3})\,6s^26p^66d^27s^17p^1$
11	$(5f^{n-2})\,6s^26p^66d^27s^2$	$(5f^{n-3})\,6s^26p^66d^47s^1$
12	$(5f^{n-2})\,6s^26p^66d^4$	$(5f^{n-3})\,6s^26p^66d^5$
13	$(5f^{n-2})\,6s^26p^66d^27s^17p^1$	$(5f^{n-3})\,6s^26p^66d^37s^2$
14	$(5f^{n-2})\,6s^26p^66d^37p^1$	$(5f^{n-3})\,6s^26p^66d^47p^1$
15	$(5f^{n-2})\,6s^26p^66d^47s^1$	$(5f^{n-3})\,6s^26p^66d^37s^17p^1$
16	$(5f^{n-2})\,6s^26p^66d^27s^27p^1$	$(5f^{n-3})\,6s^26p^66d^47s^17p^1$
17	$(5f^{n-2})\,6s^26p^66d^5$	$(5f^{n-3})\,6s^26p^66d^37s^27p^1$
18	$(5f^{n-2})\,6s^26p^66d^37s^17p^1$	$(5f^{n-3})\,6s^26p^66d^27s^27p^2$

Table A.3: Parameters of the di- ($5f^{n+1}$, n=5–13 for Pu–No; $Q = 10$) and tetravalent ($5f^{n-1}$, n=1–9 for Th–Cf; $Q = 12$) 5f-in-core PPs for actinides [29] and of the tetravalent ($4f^{n-1}$, n=1–3, 8, 9 for Ce–Nd, Tb, Dy; $Q = 12$) 4f-in-core PPs for lanthanides [28].

			Actinides[a]				Lanthanides[b]		
			$Q = 10$		$Q = 12$			$Q = 12$	
l	k	An	A_{lk}	a_{lk}	A_{lk}	a_{lk}	Ln	A_{lk}	a_{lk}
0	1	Th			101.530523	3.1878	Ce	119.423291	3.9026
0	2				-12.344959	2.5052		-2.511980	2.0907
1	1				61.875695	2.4114		76.607034	3.1582
1	2				-1.011261	1.5939		-0.430895	1.6651
2	1				27.688250	1.4416		68.864734	2.5236
2	2				1.134394	1.5308		-4.246970	2.2757
3	1				-3.474523	0.9539		-41.694411	4.6452
0	1	Pa			101.535521	3.2535	Pr	119.421511	4.0565
0	2				-12.346846	2.2728		-2.522571	2.0634
1	1				61.867648	2.5280		76.604560	3.2873
1	2				-0.997613	1.6695		-0.452512	1.6508
2	1				27.708880	1.5405		68.863419	2.6443
2	2				1.140057	1.5474		-4.248770	2.2649
3	1				-4.373234	1.1371		-42.538874	4.9055

Table A.3: (continued).

| | | | Actinides[a] | | | | Lanthanides[b] | |
| | | | $Q=10$ | | $Q=12$ | | $Q=12$ | |
l	k	An	A_{lk}	a_{lk}	A_{lk}	a_{lk}	Ln	A_{lk}	a_{lk}
3	2				1.017037	0.8389		0.725701	1.1484
0	1	U			101.510029	3.4119	Nd	119.410511	4.2588
0	2				-12.398152	2.3679		-2.481276	2.3589
1	1				61.870048	2.6658		76.581227	3.4392
1	2				-0.817262	1.9668		-0.334544	1.8026
2	1				27.680805	1.6260		68.852942	2.7769
2	2				0.948957	1.8416		-4.256800	2.3321
3	1				-5.060991	1.2775		-42.929627	5.1726
3	2				2.081285	0.8824		1.456575	1.1957
0	1	Np			101.516132	3.5983			
0	2				-12.369779	2.5430			
1	1				61.836563	2.7674			
1	2				-1.068386	1.8212			
2	1				27.712219	1.7383			
2	2				1.141543	1.5753			
3	1				-5.615963	1.4081			
3	2				3.191067	0.9264			
0	1	Pu	101.504936	4.0712	101.506983	3.6343			
0	2		-12.399453	2.5110	-12.426998	2.3087			
1	1		61.837518	3.1908	61.826324	2.8807			
1	2		-0.992683	1.8837	-1.221450	2.0087			
2	1		27.691590	2.2615	27.775505	1.8302			
2	2		1.086341	1.6683	1.193230	1.6747			
3	1		-7.478504	1.7755	-6.031720	1.5334			
3	2		4.615821	0.8123	4.344889	0.9709			
0	1	Am	101.506880	4.2423	101.523169	3.7025			
0	2		-12.380058	2.5770	-12.373584	2.1912			
1	1		61.830559	3.3341	61.837340	3.0106			
1	2		-1.023903	1.8675	-0.863600	2.0821			
2	1		27.691483	2.4192	27.756497	1.9209			
2	2		1.102097	1.6712	1.152681	1.7035			
3	1		-7.316609	1.8817	-6.299625	1.6565			
3	2		5.502241	0.8535	5.541411	1.0159			
0	1	Cm	101.504260	4.3987	101.528345	3.8235			
0	2		-12.379533	2.6117	-12.353657	2.2001			
1	1		61.828567	3.4877	61.850235	3.1239			
1	2		-0.990561	1.9095	-0.823808	1.9785			

Table A.3: (continued).

| | | | Actinides[a] | | | | Lanthanides[b] | |
| | | | $Q=10$ | | $Q=12$ | | $Q=12$ | |
l	k	An	A_{lk}	a_{lk}	A_{lk}	a_{lk}	Ln	A_{lk}	a_{lk}
2	1		27.689080	2.5736	27.739747	2.0188			
2	2		1.085596	1.7209	1.158684	1.6901			
3	1		-6.963250	1.9913	-6.408602	1.7790			
3	2		6.419253	0.8952	6.779425	1.0615			
0	1	Bk	101.506205	4.4712	101.529576	3.9455	Tb	119.384916	5.1371
0	2		-12.377891	2.4926	-12.348977	2.2115		-2.464767	2.6552
1	1		61.836181	3.6299	61.838375	3.2431		76.525459	4.1681
1	2		-0.962722	1.9044	-0.816482	2.0290		-0.169881	1.9532
2	1		27.680426	2.7315	27.755579	2.1135		68.841221	3.4016
2	2		1.069841	1.7240	1.176777	1.7604		-4.271170	2.3394
3	1		-6.404162	2.1042	-6.346459	1.9019		-37.330833	6.6747
3	2		7.366130	0.9375	8.057875	1.1076		5.141693	1.4379
0	1	Cf	101.475587	4.4749	101.529253	3.9803	Dy	119.381621	5.3029
0	2		-12.573643	2.3117	-12.353376	2.0705		-2.490913	2.6047
1	1		61.850531	3.7131	61.853084	3.3400		76.519446	4.3031
1	2		-1.337306	1.8582	-0.846896	1.9190		-0.258942	1.8863
2	1		27.675091	2.8736	27.724516	2.2084		68.839323	3.5187
2	2		1.046808	1.7588	1.150912	1.6991		-4.277393	2.3086
3	1		-5.624800	2.2206	-6.100431	2.0258		-34.391881	7.0083
3	2		8.342264	0.9803	9.375794	1.1543		5.876988	1.4877
0	1	Es	101.496988	4.7615					
0	2		-12.399519	2.5251					
1	1		61.813918	3.9146					
1	2		-1.092360	1.9517					
2	1		27.669903	3.0620					
2	2		1.009338	1.8306					
3	1		-4.610193	2.3404					
3	2		9.347159	1.0238					
0	1	Fm	101.499587	4.8149					
0	2		-12.394687	2.4000					
1	1		61.828820	4.0681					
1	2		-0.929628	1.9180					
2	1		27.661224	3.2280					
2	2		0.982986	1.8492					
3	1		-3.345067	2.4637					
3	2		10.380482	1.0679					
0	1	Md	101.499103	4.8971					

Table A.3: (continued).

			Actinides[a]					Lanthanides[b]	
			$Q=10$		$Q=12$			$Q=12$	
l	k	An	A_{lk}	a_{lk}	A_{lk}	a_{lk}	Ln	A_{lk}	a_{lk}
0	2		-12.396828	2.3291					
1	1		61.828786	4.1983					
1	2		-0.947297	1.8667					
2	1		27.652220	3.3993					
2	2		0.951381	1.8736					
3	1		-1.813704	2.5906					
3	2		11.442016	1.1127					
0	1	No	101.500914	4.7845					
0	2		-12.419287	2.0399					
1	1		61.822993	4.3161					
1	2		-0.933399	1.9569					
2	1		27.645365	3.5188					
2	2		0.932537	1.8810					
3	1		12.531620	1.1581					

[a]The parameters of the divalent 5f-in-core PPs I have already adjusted during my diploma thesis [19].
[b]The tetravalent 4f-in-core PPs were adjusted by M. Hülsen.

Table A.4: Parameters of the penta- ($5f^{n-2}$, n=2–6 for Pa–Am; $Q=13$) and hexavalent ($5f^{n-3}$, n=3–6 for U–Am; $Q=14$) 5f-in-core PPs for actinides [30].

			$Q=13$		$Q=14$	
l	k	An	A_{lk}	a_{lk}	A_{lk}	a_{lk}
0	1	Pa	101.530502	9.1888		
0	2		101.529662	3.2643		
0	3		-12.347038	2.4788		
1	1		61.875296	2.4623		
1	2		-1.012227	1.5912		
2	1		27.685976	1.4977		
2	2		1.130790	1.5368		
3	1		-2.998486	0.8632		
0	1	U	101.525621	9.3059	102.244491	9.7387
0	2		101.526474	3.3893	102.252401	3.3303
0	3		-12.352121	2.4477	-4.327744	1.4585
1	1		61.866963	2.5704	61.851913	2.4529
1	2		-0.996819	1.6769	1.172466	1.1990
2	1		27.681302	1.5852	27.320350	1.5082

Table A.4: (continued).

l	k	An	$Q=13$		$Q=14$	
			A_{lk}	a_{lk}	A_{lk}	a_{lk}
2	2		1.110842	1.5517	-0.133230	1.1772
3	1		-3.258365	0.8907	-3.194363	0.9876
3	2		1.206614	0.9566		
0	1	Np	101.530336	9.4041	102.244235	9.7490
0	2		101.515630	3.5180	102.247872	3.4993
0	3		-12.362053	2.4445	-4.325984	1.5433
1	1		61.870355	2.6571	61.843073	2.5571
1	2		-1.039633	1.4911	-1.161007	1.2810
2	1		28.204055	1.6890	27.361171	1.5838
2	2		1.629124	1.6779	-0.120114	1.1535
3	1		-3.357339	0.8907	-3.511579	1.0305
3	2		2.464953	1.0024	1.409055	1.0777
0	1	Pu	101.514401	9.6589	102.243872	9.7631
0	2		101.503391	3.7101	102.242707	3.7011
0	3		-12.380845	2.6287	-4.316857	1.7167
1	1		61.832932	2.7812	61.823982	2.6781
1	2		-1.080822	1.7817	-1.121911	1.4854
2	1		28.054819	1.7704	27.411347	1.6590
2	2		1.476696	1.7388	-0.143002	1.0969
3	1		-3.266963	0.8548	-3.754220	1.0726
3	2		3.773377	1.0487	2.874433	1.1257
0	1	Am	101.503615	9.7495	102.243881	9.7619
0	2		101.483537	3.8324	102.249453	3.8779
0	3		-12.439750	2.5969	-4.305243	1.8040
1	1		61.811778	2.8913	61.802876	2.7885
1	2		-1.231665	2.0153	-1.100244	1.6377
2	1		28.292323	1.8646	27.461971	1.7096
2	2		1.704499	1.8440	-0.296457	0.8633
3	1		-3.159664	0.8325	-3.907279	1.1111
3	2		5.130444	1.0956	4.394546	1.1743

Table A.5: Energy differences (in eV) between finite-difference AE and PP HF calculations [55] of the di- ($5f^{n+1}$, n=5–13 for Pu–No; $Q = 10$) and tetravalent ($5f^{n-1}$, n=1–9 for Th–Cf; $Q = 12$) 5f-in-core PPs for actinides [29] and of the tetravalent ($4f^{n-1}$, n=1–3, 8, 9 for Ce–Nd, Tb, Dy; $Q = 12$) 4f-in-core PPs for lanthanides [28] without f-PPs.

	Actinides						Lanthanides		
	$Q = 10$			$Q = 12$				$Q = 12$	
An	$\overline{\Delta E}^a$	ΔE^b_{max}	No.^{c}	$\overline{\Delta E}^a$	ΔE^b_{max}	No.^{c}	Ln	$\overline{\Delta E}^a$	ΔE^b_{max}	No.^{c}
Th				0.018	0.035	14	Ce	0.028	0.068	16
Pa				0.006	0.012	2	Pr	0.027	0.062	2
U				0.015	0.029	16	Nd	0.024	0.057	2
Np				0.024	0.049	16	Tb	0.018	0.035	2
Pu	0.037	0.064	3	0.029	0.062	16	Dy	0.018	0.034	14
Am	0.035	0.061	3	0.033	0.073	16				
Cm	0.033	0.058	3	0.037	0.082	16				
Bk	0.031	0.056	3	0.040	0.091	16				
Cf	0.031	0.054	3	0.042	0.095	16				
Es	0.028	0.052	3							
Fm	0.027	0.050	3							
Md	0.026	0.048	3							
No	0.026	0.045	3							

^{a}Root mean square deviation of the valence energies $\overline{\Delta E} = \sqrt{\frac{1}{n}\sum_i^n \Delta E_i^2}$, where n is the number of reference configurations and $\Delta E_i = E_i^{PP} - E_i^{AE}$.
^{b}Maximum deviation of the valence energies $\Delta E_{max} = \max(|\Delta E_i|)$.
^{c}Reference configuration, where the maximum deviation ΔE_{max} occurs (cf. Table A.1).

Table A.6: Energy differences (in eV) between finite-difference AE and PP HF calculations [55] of the penta- ($5f^{n-2}$, n=2–6 for Pa–Am; $Q = 13$) and hexavalent ($5f^{n-3}$, n=3–6 for U–Am; $Q = 14$) 5f-in-core PPs for actinides [30] without f-PPs.

	$Q = 13$			$Q = 14$		
An	$\overline{\Delta E}^a$	ΔE^b_{max}	No.^{c}	$\overline{\Delta E}^a$	ΔE^b_{max}	No.^{c}
Pa	0.021	0.043	2			
U	0.007	0.016	2	0.010	0.024	18
Np	0.007	0.013	17	0.005	0.013	4
Pu	0.017	0.038	17	0.011	0.023	4
Am	0.025	0.055	17	0.013	0.028	4

^{a}Root mean square deviation of the valence energies $\overline{\Delta E} = \sqrt{\frac{1}{n}\sum_i^n \Delta E_i^2}$, where n is the number of reference configurations and $\Delta E_i = E_i^{PP} - E_i^{AE}$.
^{b}Maximum deviation of the valence energies $\Delta E_{max} = \max(|\Delta E_i|)$.
^{c}Reference configuration, where the maximum deviation ΔE_{max} occurs (cf. Table A.2).

Table A.7: Energy differences (in eV) between finite-difference AE and PP HF calculations [55] using the f-PPs V_1 and V_2 of the di- ($5f^{n+1}$, n=5–13 for Pu–No; $Q = 10$) and tetravalent ($5f^{n-1}$, n=1–9 for Th–Cf; $Q = 12$) 5f-in-core PPs for actinides [29].

	$Q = 10$				$Q = 12$			
	V_1		V_2		V_1		V_2	
An	$\overline{\Delta E}^a$	ΔE_{max}^b	$\overline{\Delta E}^a$	ΔE_{max}^b	$\overline{\Delta E}^a$	ΔE_{max}^c	$\overline{\Delta E}^a$	ΔE_{max}^c
Th					0.077	0.099		
Pa					0.088	0.107	0.008	0.033
U					0.096	0.117	0.008	0.032
Np					0.102	0.125	0.007	0.030
Pu	0.063	0.077	0.015	0.019	0.107	0.131	0.007	0.029
Am	0.067	0.080	0.014	0.018	0.112	0.137	0.007	0.028
Cm	0.069	0.084	0.013	0.017	0.116	0.141	0.006	0.027
Bk	0.072	0.087	0.013	0.016	0.119	0.146	0.006	0.026
Cf	0.074	0.089	0.012	0.015	0.123	0.149	0.006	0.025
Es	0.076	0.091	0.011	0.014				
Fm	0.078	0.093	0.011	0.013				
Md	0.079	0.095	0.010	0.013				
No			0.009	0.012				

[a] Root mean square deviation of the valence energies $\overline{\Delta E} = \sqrt{\frac{1}{n}\sum_i^n \Delta E_i^2}$, where n is the number of reference configurations and $\Delta E_i = E_i^{PP} - E_i^{AE}$.
[b] Maximum deviation of the valence energies $\Delta E_{max} = \max(|\Delta E_i|)$ occurring for the reference configuration $5f^{n+1}8f^1$.
[c] Maximum deviation of the valence energies occurring for the reference configuration $5f^{n-1}8f^1$.

Table A.8: Energy differences (in eV) between finite-difference AE and PP HF calculations [55] using the f-PPs V_1 and V_2 of the penta- ($5f^{n-2}$, n=2–6 for Pa–Am; $Q = 13$) and hexavalent ($5f^{n-3}$, n=3–6 for U–Am; $Q = 14$) 5f-in-core PPs for actinides [30].

	$Q = 13$				$Q = 14$			
	V_1		V_2		V_1		V_2	
An	$\overline{\Delta E}^a$	ΔE_{max}^b	$\overline{\Delta E}^a$	ΔE_{max}^c	$\overline{\Delta E}^a$	ΔE_{max}^d	$\overline{\Delta E}^a$	ΔE_{max}^e
Pa	0.242	0.294						
U	0.160	0.212	0.009	0.038	0.579	0.756		
Np	0.092	0.131	0.009	0.036	0.484	0.607	0.010	0.043
Pu	0.040	0.059^c	0.008	0.035	0.395	0.471	0.010	0.041
Am	0.006	0.010	0.008	0.034	0.307	0.388	0.010	0.040

[a] Root mean square deviation of the valence energies $\overline{\Delta E} = \sqrt{\frac{1}{n}\sum_i^n \Delta E_i^2}$, where n is the number of reference configurations and $\Delta E_i = E_i^{PP} - E_i^{AE}$.
[b] Maximum deviation of the valence energies $\Delta E_{max} = \max(|\Delta E_i|)$ occurring for the reference configuration $5f^{n-2}7f^1$.
[c] Maximum deviation of the valence energies occurring for the reference configuration $5f^{n-2}8f^1$.
[d] Maximum deviation of the valence energies occurring for the reference configuration $5f^{n-3}6f^1$ and $5f^{n-3}7f^1$ for U, Np and Pu, Am, respectively.
[e] Maximum deviation of the valence energies occurring for the reference configuration $5f^{n-3}8f^1$.

Table A.9: Energy differences (in eV) between finite-difference AE and PP HF calculations [55] using the f-PPs V_1 and V_2 of the tetravalent ($4f^{n-1}$, n=1–3, 8, 9 for Ce–Nd, Tb, Dy; $Q = 12$) 4f-in-core PPs for lanthanides [28].

Ln	$\overline{\Delta E}^a$ (V_1)	ΔE^b_{max} (V_1)	$\overline{\Delta E}^a$ (V_2)	ΔE^b_{max} (V_2)
Ce	0.162	0.201		
Pr	0.150	0.189	0.022	0.028
Nd	0.139	0.176	0.021	0.027
Tb	0.092	0.121	0.016	0.021
Dy	0.083	0.111	0.015	0.020

[a] Root mean square deviation of the valence energies $\overline{\Delta E} = \sqrt{\frac{1}{n}\sum_i^n \Delta E_i^2}$, where n is the number of reference configurations and $\Delta E_i = E_i^{PP} - E_i^{AE}$.
[b] Maximum deviation of the valence energies $\Delta E_{max} = \max(|\Delta E_i|)$ occurring for the reference configuration $4f^{n-1}7f^1$.

Table A.10: Parameters of the CPPs corresponding to di- ($5f^{n+1}$, n=5–13 for Pu–No; $Q = 10$) [29], tri- ($5f^n$, n=0–14 for Ac–Lr; $Q = 11$) [54], and tetravalent ($5f^{n-1}$, n=1–9 for Th–Cf; $Q = 12$) [29] 5f-in-core PPs for actinides, respectively. Given are the dipole polarizabilities α_D (in a.u.) and cutoff parameters δ.

An	α_D ($Q=10$)	δ ($Q=10$)	α_D ($Q=11$)	δ ($Q=11$)	α_D ($Q=12$)	δ ($Q=12$)
Ac			0.8982	0.8727		
Th			1.1019	0.8296	0.7830	0.9293
Pa			1.3056	0.7865	0.9069	0.8661
U			1.5093	0.7435	1.0308	0.8028
Np			1.7130	0.7004	1.1548	0.7396
Pu	3.3726	0.7634	1.9167	0.6573	1.2787	0.6764
Am	3.7613	0.6980	2.1204	0.6142	1.4026	0.6132
Cm	4.1500	0.6326	2.3242	0.5712	1.5265	0.5499
Bk	4.5386	0.5673	2.5279	0.5281	1.6505	0.4867
Cf	4.9273	0.5019	2.7316	0.4850	1.7744	0.4235
Es	5.3159	0.4365	2.9353	0.4419		
Fm	5.7046	0.3711	3.1390	0.3988		
Md	6.0933	0.3058	3.3427	0.3558		
No	6.4819	0.2404	3.5464	0.3127		
Lr			3.7501	0.2696		

Table A.11: Reference configurations for the adjustment of the s-, p-, d-, and f-parts of the 5f-in-valence MCDHF/DCB PP for uranium ($Q = 32$) [16].

U	U$^+$	U^{+2}
$5s^25p^65d^{10}5f^56s^26p^67s^1$	$5s^25p^65d^{10}5f^56s^26p^6$	$5s^25p^65d^{10}5f^46s^26p^6$
$5s^25p^65d^{10}5f^56s^26p^67p^1$	$5s^25p^65d^{10}5f^46s^26p^66d^1$	$5s^25p^65d^{10}5f^36s^26p^66d^1$
$5s^25p^65d^{10}5f^46s^26p^66d^17s^1$	$5s^25p^65d^{10}5f^46s^26p^67s^1$	$5s^25p^65d^{10}5f^36s^26p^67s^1$
$5s^25p^65d^{10}5f^46s^26p^66d^17p^1$	$5s^25p^65d^{10}5f^46s^26p^67p^1$	$5s^25p^65d^{10}5f^36s^26p^67p^1$
$5s^25p^65d^{10}5f^46s^26p^67s^2$	$5s^25p^65d^{10}5f^36s^26p^66d^2$	$5s^25p^65d^{10}5f^26s^26p^66d^2$
$5s^25p^65d^{10}5f^46s^26p^67s^17p^1$	$5s^25p^65d^{10}5f^36s^26p^66d^17s^1$	$5s^25p^65d^{10}5f^26s^26p^66d^17s^1$
$5s^25p^65d^{10}5f^36s^26p^66d^17s^2$	$5s^25p^65d^{10}5f^36s^26p^67s^2$	$5s^25p^65d^{10}5f^26s^26p^66d^17p^1$
$5s^25p^65d^{10}5f^36s^26p^67s^27p^1$	$5s^25p^65d^{10}5f^36s^26p^67s^17p^1$	$5s^25p^65d^{10}5f^26s^26p^67s^2$
$5s^25p^65d^{10}5f^36s^26p^67s^28s^1$	$5s^25p^65d^{10}5f^36s^26p^66d^17p^1$	$5s^25p^65d^{10}5f^26s^26p^67s^17p^1$
$5s^25p^65d^{10}5f^36s^26p^67s^29s^1$	$5s^25p^65d^{10}5f^26s^26p^66d^3$	$5s^25p^65d^{10}5f^16s^26p^66d^3$
$5s^25p^65d^{10}5f^36s^26p^67s^28p^1$	$5s^25p^65d^{10}5f^26s^26p^66d^27s^1$	$5s^25p^65d^{10}5f^16s^26p^66d^27s^1$
$5s^25p^65d^{10}5f^36s^26p^67s^29p^1$	$5s^25p^65d^{10}5f^26s^26p^66d^17s^2$	$5s^25p^65d^{10}5f^16s^26p^66d^17s^2$
$5s^25p^65d^{10}5f^26s^26p^66d^27s^2$	$5s^25p^65d^{10}5f^26s^26p^66d^17s^17p^1$	$5s^25p^65d^{10}5f^16s^26p^66d^17s^17p^1$
$5s^25p^65d^{10}5f^26s^26p^66d^17s^27p^1$	$5s^25p^65d^{10}5f^26s^26p^67s^27p^1$	$5s^25p^65d^{10}5f^16s^26p^67s^27p^1$
$5s^25p^65d^{10}5f^16s^26p^66d^47s^1$	$5s^25p^65d^{10}5f^16s^26p^66d^4$	$5s^25p^65d^{10}6s^26p^66d^4$
$5s^25p^65d^{10}5f^16s^26p^66d^37s^2$	$5s^25p^65d^{10}5f^16s^26p^66d^37s^1$	$5s^25p^65d^{10}6s^26p^66d^37s^1$
$5s^25p^65d^{10}6s^26p^66d^47s^2$	$5s^25p^65d^{10}5f^16s^26p^66d^27s^2$	$5s^25p^65d^{10}6s^26p^66d^27s^2$
$5s^25p^65d^{10}6s^26p^66d^37s^27p^1$	$5s^25p^65d^{10}5f^16s^26p^66d^27s^17p^1$	$5s^25p^65d^{10}6s^26p^66d^17s^27p^1$
	$5s^25p^65d^{10}5f^16s^26p^66d^17s^27p^1$	
	$5s^25p^65d^{10}6s^26p^66d^47s^1$	
	$5s^25p^65d^{10}6s^26p^66d^27s^27p^1$	

U^{+3}	U^{+4}	U^{+5}
$5s^25p^65d^{10}5f^36s^26p^6$	$5s^25p^65d^{10}5f^26s^26p^6$	$5s^25p^65d^{10}5f^16s^26p^6$
$5s^25p^65d^{10}5f^26s^26p^66d^1$	$5s^25p^65d^{10}5f^16s^26p^66d^1$	$5s^25p^65d^{10}6s^26p^66d^1$
$5s^25p^65d^{10}5f^26s^26p^67s^1$	$5s^25p^65d^{10}5f^16s^26p^67s^1$	$5s^25p^65d^{10}6s^26p^67s^1$
$5s^25p^65d^{10}5f^26s^26p^67p^1$	$5s^25p^65d^{10}5f^16s^26p^67p^1$	$5s^25p^65d^{10}6s^26p^67p^1$
$5s^25p^65d^{10}5f^16s^26p^66d^2$	$5s^25p^65d^{10}6s^26p^66d^2$	$5s^25p^65d^{10}6s^26p^68s^1$
$5s^25p^65d^{10}5f^16s^26p^66d^17s^1$	$5s^25p^65d^{10}6s^26p^66d^17s^1$	$5s^25p^65d^{10}6s^26p^69s^1$
$5s^25p^65d^{10}5f^16s^26p^66d^17p^1$	$5s^25p^65d^{10}6s^26p^66d^17p^1$	$5s^25p^65d^{10}6s^26p^68p^1$
$5s^25p^65d^{10}5f^16s^26p^67s^2$	$5s^25p^65d^{10}6s^26p^67s^2$	$5s^25p^65d^{10}6s^26p^69p^1$
$5s^25p^65d^{10}5f^16s^26p^67s^17p^1$	$5s^25p^65d^{10}6s^26p^67s^17p^1$	$5s^25p^65d^{10}6s^26p^67d^1$
$5s^25p^65d^{10}6s^26p^66d^3$		$5s^25p^65d^{10}6s^26p^68d^1$
$5s^25p^65d^{10}6s^26p^66d^27s^1$		$5s^25p^65d^{10}6s^26p^69d^1$
$5s^25p^65d^{10}6s^26p^66d^17s^2$		$5s^25p^65d^{10}6s^26p^66f^1$
$5s^25p^65d^{10}6s^26p^67s^27p^1$		$5s^25p^65d^{10}6s^26p^67f^1$
		$5s^25p^65d^{10}6s^26p^68f^1$
		$5s^25p^65d^{10}6s^26p^69f^1$

U^{+6}	U^{+7}	U^{+7}
$5s^25p^65d^{10}5f^36s^26p^6$	$5s^15p^65d^{10}6s^26p^6$	$5s^25p^65d^{10}6s^16p^6$
	$5s^25p^55d^{10}6s^26p^6$	$5s^25p^65d^{10}6s^26p^5$
	$5s^25p^65d^96s^26p^6$	

Table A.12: Parameters[a] of the two-component 5f-in-valence MCDHF/DCB PP for uranium ($Q = 32$) [16].

l	j	k	B_{ljk}	b_{ljk}
0	1/2	1	529.53526911	16.91870874
0	1/2	2	4.27018845	3.40970576
0	1/2	3	0.09998874	0.79302733
0	1/2	4	0.00626781	0.19378381
1	1/2	1	302.80077401	13.16953414
1	3/2	1	263.93135846	10.60784728
1	1/2	2	-0.00632361	2.69049397
1	3/2	2	-0.28562472	2.08929800
1	1/2	3	0.01483880	0.54050990
1	3/2	3	-0.02478724	0.40482776
1	1/2	4	0.00246099	0.11250285
1	3/2	4	-0.00150041	0.09508873
2	3/2	1	157.14819756	9.06784123
2	5/2	1	150.34804157	8.53362678
2	3/2	2	-0.20706045	1.63646790
2	5/2	2	-0.25513195	1.54425719
2	3/2	3	-0.00021799	0.47961552
2	5/2	3	0.00806796	0.41164502
2	3/2	4	-0.00015340	0.13990510
2	5/2	4	-0.00401398	0.17494682
3	5/2	1	36.60132534	5.14746012
3	7/2	1	39.06184353	5.29241394
3	5/2	2	-0.48275111	1.05726701
3	7/2	2	-0.14760289	0.98063114
3	5/2	3	0.14197042	0.48259555
3	7/2	3	0.00404713	0.55434882
3	5/2	4	-0.00476161	0.23674544
3	7/2	4	0.00609679	0.21559852
4	7/2	1	-99.92316195	18.83643086
4	9/2	1	-96.57611061	18.74850924
4	7/2	2	-5.74243522	6.49279545
4	9/2	2	-6.01884159	6.57472519
4	7/2	3	0.10186930	2.58151924
4	9/2	3	0.10148305	2.58690949

[a]The PP was adjusted by M. Dolg.

Appendix B

Basis Sets

Table B.1: Exponents and contraction coefficients of the (7s6p5d), (7s6p5d)/[6s5p4d] {211111/21111/2111}, (7s6p5d)/[5s4p4d] {31111/3111/2111}, and (7s6p5d)/[4s3p3d] {3211/321/311} as well as of the (6s5p4d), (6s5p4d)/[5s4p3d] {21111/2111/211}, and (6s5p4d)/[4s3p3d] {2211/221/211} GTO valence basis sets for the divalent ($5f^{n+1}$, n=5–13 for Pu–No) 5f-in-core PPs for actinides [29]. Additionally, the two f and one g polarization functions are given.

An		(7s6p5d)/	[6s5p4d]	[5s4p4d]	[4s3p3d]	(6s5p4d)/	[5s4p3d]	[4s3p3d]
Pu	s	7.099797	-0.102293	-0.102293	-0.102293	4.071880	0.476802	0.476802
		4.733198	0.620330	0.620330	0.620330	2.714587	-0.997050	-0.997050
		2.974894	1.000000	-0.991321	-0.991321	0.640897	1.000000	0.807075
		0.568363	1.000000	1.000000	0.892671	0.277613	1.000000	0.524547
		0.249491	1.000000	1.000000	0.393614	0.062021	1.000000	1.000000
		0.063629	1.000000	1.000000	1.000000	0.027540	1.000000	1.000000
		0.027857	1.000000	1.000000	1.000000			
	p	3.855776	0.107344	0.107344	0.107344	4.018228	0.099782	0.099782
		2.582683	-0.286647	-0.286647	-0.286647	2.670701	-0.262835	-0.262835
		0.680070	1.000000	0.421465	0.421465	0.566548	1.000000	0.632849
		0.329090	1.000000	1.000000	0.520889	0.221476	1.000000	0.476800
		0.154720	1.000000	1.000000	0.204110	0.070541	1.000000	1.000000
		0.064100	1.000000	1.000000	1.000000			
	d	2.010101	0.018281	0.018281	0.018281	2.483170	-0.011281	-0.011281
		0.589728	-0.168424	-0.168424	-0.168424	0.474833	0.246988	0.246988
		0.213274	1.000000	1.000000	-0.335810	0.136942	1.000000	1.000000
		0.077409	1.000000	1.000000	1.000000	0.041147	1.000000	1.000000
		0.027972	1.000000	1.000000	1.000000			
	f	0.970						
		0.403						

Table B.1: (continued).

An		(7s6p5d)/	[6s5p4d]	[5s4p4d]	[4s3p3d]	(6s5p4d)/	[5s4p3d]	[4s3p3d]
	g	0.853						
Am	s	7.311443	-0.094661	-0.094661	-0.094661	4.332225	0.459342	0.459342
		4.874295	0.632757	0.632757	0.632757	2.888150	-0.958347	-0.958347
		3.150528	1.000000	-1.001778	-1.001778	0.649477	1.000000	0.823474
		0.594258	1.000000	1.000000	0.884793	0.281157	1.000000	0.491287
		0.259319	1.000000	1.000000	0.395678	0.064995	1.000000	1.000000
		0.065370	1.000000	1.000000	1.000000	0.028758	1.000000	1.000000
		0.028490	1.000000	1.000000	1.000000			
	p	3.780929	0.173901	0.173901	0.173901	4.279281	0.096429	0.096429
		2.879710	-0.342251	-0.342251	-0.342251	2.852854	-0.250887	-0.250887
		0.694187	1.000000	0.435554	0.435554	0.587520	1.000000	0.632659
		0.333442	1.000000	1.000000	0.509664	0.228009	1.000000	0.473504
		0.157034	1.000000	1.000000	0.195792	0.072149	1.000000	1.000000
		0.064705	1.000000	1.000000	1.000000			
	d	2.184394	0.015557	0.015557	0.015557	2.775368	-0.009052	-0.009052
		0.625243	-0.162404	-0.162404	-0.162404	0.499709	0.241657	0.241657
		0.224275	1.000000	1.000000	-0.327844	0.141397	1.000000	1.000000
		0.080187	1.000000	1.000000	1.000000	0.041770	1.000000	1.000000
		0.028490	1.000000	1.000000	1.000000			
	f	0.991						
		0.415						
	g	0.888						
Cm	s	7.493885	-0.088834	-0.088834	-0.088834	4.528355	0.449640	0.449640
		4.995923	0.660351	0.660351	0.660351	3.018903	-0.944447	-0.944447
		3.330616	1.000000	-1.027970	-1.027970	0.685936	1.000000	0.810533
		0.621922	1.000000	1.000000	0.877387	0.294443	1.000000	0.502905
		0.269763	1.000000	1.000000	0.399151	0.066637	1.000000	1.000000
		0.067171	1.000000	1.000000	1.000000	0.029364	1.000000	1.000000
		0.029131	1.000000	1.000000	1.000000			
	p	4.057822	0.144139	0.144139	0.144139	4.485808	0.090897	0.090897
		3.002285	-0.305917	-0.305917	-0.305917	2.990538	-0.240087	-0.240087
		0.724461	1.000000	0.433570	0.433570	0.615619	1.000000	0.625974
		0.344967	1.000000	1.000000	0.510082	0.236723	1.000000	0.479166
		0.161425	1.000000	1.000000	0.195549	0.073715	1.000000	1.000000
		0.066064	1.000000	1.000000	1.000000			
	d	2.390505	0.012868	0.012868	0.012868	3.092587	0.007200	0.007200
		0.654420	-0.159252	-0.159252	-0.159252	0.524586	-0.236158	-0.236158
		0.231377	1.000000	1.000000	-0.323373	0.145602	1.000000	1.000000
		0.081656	1.000000	1.000000	1.000000	0.042301	1.000000	1.000000

Table B.1: (continued).

An		(7s6p5d)/	[6s5p4d]	[5s4p4d]	[4s3p3d]	(6s5p4d)/	[5s4p3d]	[4s3p3d]
		0.028627	1.000000	1.000000	1.000000			
	f	1.012						
		0.426						
	g	0.922						
Bk	s	7.750942	-0.090316	-0.090316	-0.090316	4.676718	0.455176	0.455176
		5.167295	0.668340	0.668340	0.668340	3.117812	-0.958692	-0.958692
		3.444863	1.000000	-1.040472	-1.040472	0.729488	1.000000	0.805712
		0.658979	1.000000	1.000000	0.870237	0.309324	1.000000	0.516962
		0.283798	1.000000	1.000000	0.412823	0.068470	1.000000	1.000000
		0.069000	1.000000	1.000000	1.000000	0.030076	1.000000	1.000000
		0.029797	1.000000	1.000000	1.000000			
	p	4.354756	0.116869	0.116869	0.116869	4.717032	0.086946	0.086946
		3.100354	-0.273636	-0.273636	-0.273636	3.144688	-0.230293	-0.230293
		0.756531	1.000000	0.428620	0.428620	0.639243	1.000000	0.622801
		0.357799	1.000000	1.000000	0.511905	0.243939	1.000000	0.480620
		0.166290	1.000000	1.000000	0.197705	0.075148	1.000000	1.000000
		0.067444	1.000000	1.000000	1.000000			
	d	2.611617	-0.010546	-0.010546	-0.010546	3.445159	0.005629	0.005629
		0.682940	0.155758	0.155758	0.155758	0.548544	-0.230468	-0.230468
		0.237969	1.000000	1.000000	0.318973	0.149407	1.000000	1.000000
		0.082941	1.000000	1.000000	1.000000	0.042719	1.000000	1.000000
		0.028700	1.000000	1.000000	1.000000			
	f	1.033						
		0.438						
	g	0.957						
Cf	s	7.900351	-0.095925	-0.095925	-0.095925	4.801003	0.470520	0.470520
		5.266901	0.698194	0.698194	0.698194	3.200668	-0.987285	-0.987285
		3.511267	1.000000	-1.083094	-1.083094	0.762881	1.000000	0.824487
		0.704620	1.000000	1.000000	0.866359	0.320322	1.000000	0.510566
		0.300880	1.000000	1.000000	0.433340	0.070318	1.000000	1.000000
		0.070639	1.000000	1.000000	1.000000	0.030736	1.000000	1.000000
		0.030401	1.000000	1.000000	1.000000			
	p	4.559154	0.106903	0.106903	0.106903	4.901994	0.084879	0.084879
		3.193051	-0.261125	-0.261125	-0.261125	3.267996	-0.224879	-0.224879
		0.787977	1.000000	0.424217	0.424217	0.661926	1.000000	0.620911
		0.370480	1.000000	1.000000	0.514184	0.250806	1.000000	0.482220
		0.171080	1.000000	1.000000	0.200190	0.076488	1.000000	1.000000
		0.068840	1.000000	1.000000	1.000000			
	d	2.827024	-0.008751	-0.008751	-0.008751	3.795928	-0.004436	-0.004436

Table B.1: (continued).

An		(7s6p5d)/	[6s5p4d]	[5s4p4d]	[4s3p3d]	(6s5p4d)/	[5s4p3d]	[4s3p3d]
		0.709159	0.152235	0.152235	0.152235	0.570004	0.224995	0.224995
		0.243598	1.000000	1.000000	0.314796	0.152478	1.000000	1.000000
		0.083937	1.000000	1.000000	1.000000	0.042965	1.000000	1.000000
		0.028682	1.000000	1.000000	1.000000			
	f	1.054						
		0.450						
	g	0.991						
Es	s	8.316404	-0.072460	-0.072460	-0.072460	5.109894	0.439540	0.439540
		5.544270	0.612414	0.612414	0.612414	3.406596	-0.929303	-0.929303
		3.696180	1.000000	-0.997598	-0.997598	0.785894	1.000000	0.808788
		0.726950	1.000000	1.000000	0.854744	0.329544	1.000000	0.505793
		0.309005	1.000000	1.000000	0.428821	0.072513	1.000000	1.000000
		0.072612	1.000000	1.000000	1.000000	0.031496	1.000000	1.000000
		0.031057	1.000000	1.000000	1.000000			
	p	4.994225	0.082035	0.082035	0.082035	5.212368	0.079729	0.079729
		3.317480	-0.228903	-0.228903	-0.228903	3.474912	-0.211715	-0.211715
		0.823677	1.000000	0.419854	0.419854	0.688814	1.000000	0.616651
		0.384100	1.000000	1.000000	0.515219	0.258864	1.000000	0.483658
		0.176126	1.000000	1.000000	0.201399	0.078105	1.000000	1.000000
		0.070304	1.000000	1.000000	1.000000			
	d	3.136682	0.006816	0.006816	0.006816	4.311882	-0.003183	-0.003183
		0.744669	-0.148165	-0.148165	-0.148165	0.598759	0.218808	0.218808
		0.251518	1.000000	1.000000	-0.309563	0.156688	1.000000	1.000000
		0.085366	1.000000	1.000000	1.000000	0.043360	1.000000	1.000000
		0.028730	1.000000	1.000000	1.000000			
	f	1.075						
		0.461						
	g	1.025						
Fm	s	8.507957	-0.074200	-0.074200	-0.074200	5.273309	0.447506	0.447506
		5.671971	0.626126	0.626126	0.626126	3.515539	-0.942576	-0.942576
		3.781314	1.000000	-1.021500	-1.021500	0.813477	1.000000	0.827112
		0.769324	1.000000	1.000000	0.854495	0.338844	1.000000	0.492972
		0.324360	1.000000	1.000000	0.440449	0.074616	1.000000	1.000000
		0.074403	1.000000	1.000000	1.000000	0.032200	1.000000	1.000000
		0.031680	1.000000	1.000000	1.000000			
	p	5.365571	0.067241	0.067241	0.067241	5.456639	0.076576	0.076576
		3.405427	-0.210012	-0.210012	-0.210012	3.637760	-0.203465	-0.203465
		0.857327	1.000000	0.414129	0.414129	0.711721	1.000000	0.614045
		0.397544	1.000000	1.000000	0.517271	0.265625	1.000000	0.484869

Table B.1: (continued).

An		(7s6p5d)/	[6s5p4d]	[5s4p4d]	[4s3p3d]	(6s5p4d)/	[5s4p3d]	[4s3p3d]
		0.181085	1.000000	1.000000	0.204512	0.079348	1.000000	1.000000
		0.071676	1.000000	1.000000	1.000000			
	d	3.424275	-0.005412	-0.005412	-0.005412	4.831599	-0.002290	-0.002290
		0.774861	0.143988	0.143988	0.143988	0.621976	0.212968	0.212968
		0.257721	1.000000	1.000000	0.304813	0.159633	1.000000	1.000000
		0.086384	1.000000	1.000000	1.000000	0.043510	1.000000	1.000000
		0.028678	1.000000	1.000000	1.000000			
	f	1.096						
		0.473						
	g	1.060						
Md	s	8.715120	-0.072687	-0.072687	-0.072687	5.412826	0.452715	0.452715
		5.810080	0.627554	0.627554	0.627554	3.608551	-0.956257	-0.956257
		3.873387	1.000000	-1.032597	-1.032597	0.853816	1.000000	0.832209
		0.808963	1.000000	1.000000	0.856302	0.352783	1.000000	0.496699
		0.338605	1.000000	1.000000	0.447086	0.076347	1.000000	1.000000
		0.076209	1.000000	1.000000	1.000000	0.032781	1.000000	1.000000
		0.032287	1.000000	1.000000	1.000000			
	p	5.703359	0.058196	0.058196	0.058196	5.736566	0.074816	0.074816
		3.497092	-0.197696	-0.197696	-0.197696	3.824377	-0.196416	-0.196416
		0.891197	1.000000	0.409504	0.409504	0.732031	1.000000	0.613880
		0.410946	1.000000	1.000000	0.519074	0.271610	1.000000	0.483373
		0.186013	1.000000	1.000000	0.207291	0.080629	1.000000	1.000000
		0.072898	1.000000	1.000000	1.000000			
	d	3.756518	-0.004198	-0.004198	-0.004198	5.444997	-0.001557	-0.001557
		0.805555	0.139833	0.139833	0.139833	0.645182	0.207128	0.207128
		0.263620	1.000000	1.000000	0.300115	0.162321	1.000000	1.000000
		0.087250	1.000000	1.000000	1.000000	0.043576	1.000000	1.000000
		0.028568	1.000000	1.000000	1.000000			
	f	1.116						
		0.483						
	g	1.093						
No	s	8.722728	-0.098821	-0.098821	-0.098821	5.288353	0.497653	0.497653
		5.815152	0.729342	0.729342	0.729342	3.525569	-1.059146	-1.059146
		3.876768	1.000000	-1.146147	-1.146147	0.923425	1.000000	0.859944
		0.848505	1.000000	1.000000	0.891803	0.375702	1.000000	0.519828
		0.352884	1.000000	1.000000	0.442629	0.076856	1.000000	1.000000
		0.077637	1.000000	1.000000	1.000000	0.032970	1.000000	1.000000
		0.032761	1.000000	1.000000	1.000000			
	p	6.017508	0.051013	0.051013	0.051013	5.937963	0.072491	0.072491

Table B.1: (continued).

An		(7s6p5d)/	[6s5p4d]	[5s4p4d]	[4s3p3d]	(6s5p4d)/	[5s4p3d]	[4s3p3d]
		3.572885	-0.187133	-0.187133	-0.187133	3.958642	-0.190189	-0.190189
		0.917330	1.000000	0.405549	0.405549	0.749963	1.000000	0.611629
		0.421114	1.000000	1.000000	0.520228	0.276729	1.000000	0.484545
		0.189622	1.000000	1.000000	0.209569	0.081648	1.000000	1.000000
		0.073885	1.000000	1.000000	1.000000			
	d	4.009185	-0.003444	-0.003444	-0.003444	5.949534	-0.001120	-0.001120
		0.825714	0.135934	0.135934	0.135934	0.658990	0.201963	0.201963
		0.266486	1.000000	1.000000	0.296459	0.163206	1.000000	1.000000
		0.087435	1.000000	1.000000	1.000000	0.043326	1.000000	1.000000
		0.028305	1.000000	1.000000	1.000000			
	f	1.125						
		0.489						
	g	1.119						

Table B.2: Exponents and contraction coefficients of the (7s6p5d), (7s6p5d)/[6s5p4d] {211111/21111/2111}, (7s6p5d)/[5s4p4d] {31111/3111/2111}, and (7s6p5d)/[4s3p3d] {3211/321/311} as well as of the (6s5p4d), (6s5p4d)/[5s4p3d] {21111/2111/211}, and (6s5p4d)/[4s3p3d] {2211/221/211} GTO valence basis sets for the tetravalent ($5f^{n-1}$, n=1–9 for Th–Cf) 5f-in-core PPs for actinides [29]. Additionally, the two f and one g polarization functions are given.

An		(7s6p5d)/	[6s5p4d]	[5s4p3d]	[4s3p3d]	(6s5p4d)/	[5s4p3d]	[4s3p3d]
Th	s	5.468457	-0.158387	-0.158387	-0.158387	2.944324	0.651902	0.651902
		3.645638	0.798282	0.798282	0.798282	1.962883	-1.370731	-1.370731
		2.212054	1.000000	-1.233689	-1.233689	0.621764	1.000000	0.826350
		0.483676	1.000000	1.000000	1.022242	0.285263	1.000000	0.663031
		0.231972	1.000000	1.000000	0.344893	0.070702	1.000000	1.000000
		0.068719	1.000000	1.000000	1.000000	0.028364	1.000000	1.000000
		0.027687	1.000000	1.000000	1.000000			
	p	2.809462	0.181883	0.181883	0.181883	2.973305	0.160561	0.160561
		1.872974	-0.497845	-0.497845	-0.497845	1.982204	-0.411147	-0.411147
		0.839119	1.000000	0.261411	0.261411	0.521207	1.000000	0.696137
		0.406170	1.000000	1.000000	0.677551	0.228925	1.000000	0.452368
		0.191253	1.000000	1.000000	0.287014	0.083835	1.000000	1.000000
		0.080093	1.000000	1.000000	1.000000			
	d	1.064148	-0.125687	-0.125687	-0.125687	1.286162	-0.055464	-0.055464
		0.709432	0.142746	0.142746	0.142746	0.387118	0.417875	0.417875
		0.349245	1.000000	1.000000	0.361280	0.150000	1.000000	1.000000

Table B.2: (continued).

An		(7s6p5d)/	[6s5p4d]	[5s4p4d]	[4s3p3d]	(6s5p4d)/	[5s4p3d]	[4s3p3d]
		0.150000	1.000000	1.000000	1.000000	0.054712	1.000000	1.000000
		0.055361	1.000000	1.000000	1.000000			
	f	0.912						
		0.278						
	g	0.703						
Pa	s	5.573824	-0.170750	-0.170750	-0.170750	3.123244	0.644641	0.644641
		3.715801	0.945438	0.945438	0.945438	2.082163	-1.353501	-1.353501
		2.409281	1.000000	-1.360165	-1.360165	0.650725	1.000000	0.833832
		0.510683	1.000000	1.000000	1.012922	0.296686	1.000000	0.648425
		0.243315	1.000000	1.000000	0.348474	0.073366	1.000000	1.000000
		0.071768	1.000000	1.000000	1.000000	0.029225	1.000000	1.000000
		0.028658	1.000000	1.000000	1.000000			
	p	2.988180	0.168687	0.168687	0.168687	3.165163	0.151755	0.151755
		1.992120	-0.460546	-0.460546	-0.460546	2.110109	-0.389044	-0.389044
		0.819483	1.000000	0.280433	0.280433	0.539872	1.000000	0.693554
		0.410720	1.000000	1.000000	0.652335	0.235620	1.000000	0.446835
		0.194468	1.000000	1.000000	0.275281	0.086352	1.000000	1.000000
		0.081610	1.000000	1.000000	1.000000			
	d	1.117356	-0.130854	-0.130854	-0.130854	1.442878	-0.046302	-0.046302
		0.744904	0.175267	0.175267	0.175267	0.401088	0.430372	0.430372
		0.344821	1.000000	1.000000	0.375672	0.152869	1.000000	1.000000
		0.150000	1.000000	1.000000	1.000000	0.055526	1.000000	1.000000
		0.056053	1.000000	1.000000	1.000000			
	f	0.901						
		0.278						
	g	0.735						
U	s	5.800429	-0.156268	-0.156268	-0.156268	3.339706	0.612438	0.612438
		3.866953	0.939036	0.939036	0.939036	2.226471	-1.282943	-1.282943
		2.560038	1.000000	-1.353120	-1.353120	0.655457	1.000000	0.845180
		0.535133	1.000000	1.000000	1.000850	0.302082	1.000000	0.603980
		0.253494	1.000000	1.000000	0.350321	0.075644	1.000000	1.000000
		0.074302	1.000000	1.000000	1.000000	0.029962	1.000000	1.000000
		0.029483	1.000000	1.000000	1.000000			
	p	3.204743	0.154087	0.154087	0.154087	3.363605	0.142637	0.142637
		2.136495	-0.417094	-0.417094	-0.417094	2.242404	-0.367635	-0.367635
		0.765451	1.000000	0.335252	0.335252	0.561215	1.000000	0.688528
		0.401216	1.000000	1.000000	0.602078	0.243270	1.000000	0.444956
		0.193892	1.000000	1.000000	0.249671	0.088701	1.000000	1.000000
		0.081989	1.000000	1.000000	1.000000			

Table B.2: (continued).

An		(7s6p5d)/	[6s5p4d]	[5s4p4d]	[4s3p3d]	(6s5p4d)/	[5s4p3d]	[4s3p3d]
	d	1.173224	-0.130632	-0.130632	-0.130632	1.576705	-0.040131	-0.040131
		0.782149	0.194476	0.194476	0.194476	0.423677	0.428020	0.428020
		0.344966	1.000000	1.000000	0.390543	0.159540	1.000000	1.000000
		0.150000	1.000000	1.000000	1.000000	0.057249	1.000000	1.000000
		0.056544	1.000000	1.000000	1.000000			
	f	0.901						
		0.282						
	g	0.766						
Np	s	6.058466	-0.133771	-0.133771	-0.133771	3.566512	0.578527	0.578527
		4.038977	0.878468	0.878468	0.878468	2.377675	-1.210922	-1.210922
		2.692652	1.000000	-1.298891	-1.298891	0.658720	1.000000	0.856603
		0.558954	1.000000	1.000000	0.987907	0.306245	1.000000	0.560154
		0.263475	1.000000	1.000000	0.352526	0.077795	1.000000	1.000000
		0.076690	1.000000	1.000000	1.000000	0.030667	1.000000	1.000000
		0.030272	1.000000	1.000000	1.000000			
	p	3.399345	0.144804	0.144804	0.144804	3.550743	0.136005	0.136005
		2.266230	-0.390894	-0.390894	-0.390894	2.367162	-0.351776	-0.351776
		0.746834	1.000000	0.388820	0.388820	0.584259	1.000000	0.684071
		0.392356	1.000000	1.000000	0.564587	0.251456	1.000000	0.444991
		0.192425	1.000000	1.000000	0.223894	0.090964	1.000000	1.000000
		0.082260	1.000000	1.000000	1.000000			
	d	1.222978	-0.131684	-0.131684	-0.131684	1.712762	-0.035549	-0.035549
		0.815319	0.212357	0.212357	0.212357	0.445930	0.424961	0.424961
		0.345970	1.000000	1.000000	0.401777	0.165909	1.000000	1.000000
		0.150000	1.000000	1.000000	1.000000	0.058808	1.000000	1.000000
		0.056895	1.000000	1.000000	1.000000			
	f	0.914						
		0.287						
	g	0.796						
Pu	s	6.315949	-0.138929	-0.138929	-0.138929	3.727244	0.584544	0.584544
		4.210633	0.897498	0.897498	0.897498	2.484829	-1.220632	-1.220632
		2.807088	1.000000	-1.319982	-1.319982	0.691633	1.000000	0.867354
		0.596166	1.000000	1.000000	0.974098	0.317994	1.000000	0.554588
		0.279706	1.000000	1.000000	0.374308	0.080700	1.000000	1.000000
		0.080209	1.000000	1.000000	1.000000	0.031608	1.000000	1.000000
		0.031383	1.000000	1.000000	1.000000			
	p	3.608300	0.135678	0.135678	0.135678	3.737805	0.129243	0.129243
		2.405514	-0.364063	-0.364063	-0.364063	2.491870	-0.335448	-0.335448
		0.727022	1.000000	0.446372	0.446372	0.605052	1.000000	0.679906

Table B.2: (continued).

An		(7s6p5d)/	[6s5p4d]	[5s4p4d]	[4s3p3d]	(6s5p4d)/	[5s4p3d]	[4s3p3d]
		0.377916	1.000000	1.000000	0.527161	0.258760	1.000000	0.444262
		0.188132	1.000000	1.000000	0.192182	0.093075	1.000000	1.000000
		0.085303	1.000000	1.000000	1.000000			
	d	1.278502	-0.129431	-0.129431	-0.129431	1.846905	-0.031477	-0.031477
		0.852335	0.220327	0.220327	0.220327	0.466803	0.421681	0.421681
		0.349884	1.000000	1.000000	0.413103	0.171736	1.000000	1.000000
		0.150000	1.000000	1.000000	1.000000	0.060170	1.000000	1.000000
		0.057057	1.000000	1.000000	1.000000			
	f	0.921						
		0.292						
	g	0.826						
Am	s	6.556178	-0.139519	-0.139519	-0.139519	3.865468	0.590748	0.590748
		4.370785	0.903028	0.903028	0.903028	2.576977	-1.237671	-1.237671
		2.913857	1.000000	-1.331147	-1.331147	0.736582	1.000000	0.865220
		0.632669	1.000000	1.000000	0.964587	0.333784	1.000000	0.568803
		0.295028	1.000000	1.000000	0.391188	0.083607	1.000000	1.000000
		0.083492	1.000000	1.000000	1.000000	0.032566	1.000000	1.000000
		0.032429	1.000000	1.000000	1.000000			
	p	3.792452	0.129369	0.129369	0.129369	3.928879	0.123548	0.123548
		2.528297	-0.348048	-0.348048	-0.348048	2.619253	-0.321530	-0.321530
		0.746079	1.000000	0.453251	0.453251	0.626290	1.000000	0.675916
		0.385401	1.000000	1.000000	0.519179	0.266187	1.000000	0.444135
		0.191654	1.000000	1.000000	0.188158	0.095135	1.000000	1.000000
		0.086887	1.000000	1.000000	1.000000			
	d	1.321431	-0.127391	-0.127391	-0.127391	1.986296	-0.027494	-0.027494
		0.880954	0.230271	0.230271	0.230271	0.487309	0.417920	0.417920
		0.352730	1.000000	1.000000	0.420166	0.177323	1.000000	1.000000
		0.150000	1.000000	1.000000	1.000000	0.061410	1.000000	1.000000
		0.057139	1.000000	1.000000	1.000000			
	f	0.927						
		0.296						
	g	0.856						
Cm	s	6.810270	-0.128364	-0.128364	-0.128364	4.050860	0.579446	0.579446
		4.540180	0.868588	0.868588	0.868588	2.700573	-1.214836	-1.214836
		3.026786	1.000000	-1.304224	-1.304224	0.761379	1.000000	0.868963
		0.664331	1.000000	1.000000	0.955359	0.343839	1.000000	0.556799
		0.308011	1.000000	1.000000	0.399880	0.086291	1.000000	1.000000
		0.086441	1.000000	1.000000	1.000000	0.033449	1.000000	1.000000
		0.033387	1.000000	1.000000	1.000000			

Table B.2: (continued).

An		(7s6p5d)/	[6s5p4d]	[5s4p4d]	[4s3p3d]	(6s5p4d)/	[5s4p3d]	[4s3p3d]
	p	3.978111	0.123962	0.123962	0.123962	4.121704	0.118650	0.118650
		2.652057	-0.334043	-0.334043	-0.334043	2.747803	-0.309258	-0.309258
		0.766302	1.000000	0.461421	0.461421	0.648847	1.000000	0.672198
		0.392221	1.000000	1.000000	0.512720	0.274020	1.000000	0.444565
		0.194590	1.000000	1.000000	0.182583	0.097182	1.000000	1.000000
		0.088251	1.000000	1.000000	1.000000			
	d	1.368744	-0.124063	-0.124063	-0.124063	2.136760	-0.023835	-0.023835
		0.912496	0.236201	0.236201	0.236201	0.508190	0.413613	0.413613
		0.356888	1.000000	1.000000	0.426811	0.182870	1.000000	1.000000
		0.150000	1.000000	1.000000	1.000000	0.062576	1.000000	1.000000
		0.057095	1.000000	1.000000	1.000000			
	f	0.934						
		0.300						
	g	0.886						
Bk	s	7.059960	-0.116655	-0.116655	-0.116655	4.239436	0.568745	0.568745
		4.706640	0.833205	0.833205	0.833205	2.826290	-1.193028	-1.193028
		3.137760	1.000000	-1.277292	-1.277292	0.786266	1.000000	0.872945
		0.696913	1.000000	1.000000	0.946918	0.353868	1.000000	0.544984
		0.321211	1.000000	1.000000	0.408180	0.089061	1.000000	1.000000
		0.089466	1.000000	1.000000	1.000000	0.034363	1.000000	1.000000
		0.034370	1.000000	1.000000	1.000000			
	p	4.171384	0.118510	0.118510	0.118510	4.323449	0.113656	0.113656
		2.780903	-0.319823	-0.319823	-0.319823	2.882300	-0.296579	-0.296579
		0.787529	1.000000	0.466936	0.466936	0.671303	1.000000	0.668507
		0.399931	1.000000	1.000000	0.507244	0.281759	1.000000	0.444802
		0.197916	1.000000	1.000000	0.178442	0.099188	1.000000	1.000000
		0.089790	1.000000	1.000000	1.000000			
	d	1.491810	-0.088617	-0.088617	-0.088617	2.292864	-0.020690	-0.020690
		0.886379	0.218080	0.218080	0.218080	0.529017	0.409081	0.409081
		0.356961	1.000000	1.000000	0.424780	0.188269	1.000000	1.000000
		0.150000	1.000000	1.000000	1.000000	0.063656	1.000000	1.000000
		0.057060	1.000000	1.000000	1.000000			
	f	0.940						
		0.305						
	g	0.917						
Cf	s	7.229916	-0.121318	-0.121318	-0.121318	4.325744	0.587566	0.587566
		4.819944	0.859321	0.859321	0.859321	2.883828	-1.239227	-1.239227
		3.213296	1.000000	-1.316454	-1.316454	0.841984	1.000000	0.880211
		0.740239	1.000000	1.000000	0.947552	0.372597	1.000000	0.564639

Table B.2: (continued).

An		(7s6p5d)/	[6s5p4d]	[5s4p4d]	[4s3p3d]	(6s5p4d)/	[5s4p3d]	[4s3p3d]
		0.338437	1.000000	1.000000	0.424727	0.091868	1.000000	1.000000
		0.092748	1.000000	1.000000	1.000000	0.035326	1.000000	1.000000
		0.035446	1.000000	1.000000	1.000000			
	p	4.244511	0.131441	0.131441	0.131441	4.498179	0.110216	0.110216
		2.935243	-0.325240	-0.325240	-0.325240	2.998786	-0.287607	-0.287607
		0.804472	1.000000	0.473709	0.473709	0.692251	1.000000	0.665588
		0.406082	1.000000	1.000000	0.500604	0.288910	1.000000	0.445449
		0.200757	1.000000	1.000000	0.174594	0.101055	1.000000	1.000000
		0.091034	1.000000	1.000000	1.000000			
	d	1.625009	-0.065305	-0.065305	-0.065305	2.451632	-0.017746	-0.017746
		0.861659	0.212309	0.212309	0.212309	0.548467	0.404417	0.404417
		0.356030	1.000000	1.000000	0.420208	0.193196	1.000000	1.000000
		0.150000	1.000000	1.000000	1.000000	0.064572	1.000000	1.000000
		0.056986	1.000000	1.000000	1.000000			
	f	0.947						
		0.309						
	g	0.947						

Table B.3: Exponents and contraction coefficients of the (7s6p5d), (7s6p5d)/[6s5p4d] {211111/21111/2111}, (7s6p5d)/[5s4p4d] {31111/3111/2111}, and (7s6p5d)/[4s3p3d] {3211/321/311} as well as of the (6s5p4d), (6s5p4d)/[5s4p3d] {21111/2111/211}, and (6s5p4d)/[4s3p3d] {2211/221/211} GTO valence basis sets for the pentavalent ($5f^{n-2}$, n=2–6 for Pa–Am) 5f-in-core PPs for actinides [30]. Additionally, the two f and one g polarization functions are given.

An		(7s6p5d)/	[6s5p4d]	[5s4p4d]	[4s3p3d]	(6s5p4d)/	[5s4p3d]	[4s3p3d]
Pa	s	6.132986	-0.143519	-0.143519	-0.143519	3.127304	0.681823	0.681823
		4.088646	0.663319	0.663319	0.663319	2.084855	-1.446673	-1.446673
		2.254696	1.000000	-1.159002	-1.159002	0.688157	1.000000	0.822992
		0.539877	1.000000	1.000000	1.021778	0.324979	1.000000	0.693446
		0.267869	1.000000	1.000000	0.374587	0.079281	1.000000	1.000000
		0.076827	1.000000	1.000000	1.000000	0.031696	1.000000	1.000000
		0.030796	1.000000	1.000000	1.000000			
	p	2.956749	0.187513	0.187513	0.187513	3.130956	0.164099	0.164099
		1.971166	-0.520341	-0.520341	-0.520341	2.087304	-0.423304	-0.423304
		0.944512	1.000000	0.249741	0.249741	0.573041	1.000000	0.690540
		0.460239	1.000000	1.000000	0.676024	0.262503	1.000000	0.448935
		0.223701	1.000000	1.000000	0.304717	0.107025	1.000000	1.000000

Table B.3: (continued).

An		(7s6p5d)/	[6s5p4d]	[5s4p3d]	[4s3p3d]	(6s5p4d)/	[5s4p3d]	[4s3p3d]
		0.098445	1.000000	1.000000	1.000000			
	d	1.067194	-0.170115	-0.170115	-0.170115	1.350788	-0.058675	-0.058675
		0.711463	0.276023	0.276023	0.276023	0.443728	0.421707	0.421707
		0.313444	1.000000	1.000000	0.415618	0.182537	1.000000	1.000000
		0.150000	1.000000	1.000000	1.000000	0.068750	1.000000	1.000000
		0.061304	1.000000	1.000000	1.000000			
	f	1.017						
		0.306						
	g	0.734						
U	s	6.504343	-0.137055	-0.137055	-0.137055	3.314919	0.667464	0.667464
		4.336140	0.637144	0.637144	0.637144	2.209946	-1.417182	-1.417182
		2.377876	1.000000	-1.130446	-1.130446	0.720259	1.000000	0.815427
		0.568155	1.000000	1.000000	1.009197	0.338400	1.000000	0.688409
		0.280307	1.000000	1.000000	0.380751	0.082149	1.000000	1.000000
		0.079813	1.000000	1.000000	1.000000	0.032626	1.000000	1.000000
		0.031760	1.000000	1.000000	1.000000			
	p	3.160687	0.170241	0.170241	0.170241	3.304717	0.155767	0.155767
		2.107125	-0.462405	-0.462405	-0.462405	2.203145	-0.403076	-0.403076
		0.835051	1.000000	0.307285	0.307285	0.593567	1.000000	0.685730
		0.444334	1.000000	1.000000	0.618072	0.270298	1.000000	0.446659
		0.221145	1.000000	1.000000	0.273815	0.109845	1.000000	1.000000
		0.097624	1.000000	1.000000	1.000000			
	d	1.110195	-0.168355	-0.168355	-0.168355	1.460411	-0.051845	-0.051845
		0.740130	0.292639	0.292639	0.292639	0.466334	0.420431	0.420431
		0.316760	1.000000	1.000000	0.427352	0.190023	1.000000	1.000000
		0.150000	1.000000	1.000000	1.000000	0.070892	1.000000	1.000000
		0.061693	1.000000	1.000000	1.000000			
	f	0.975						
		0.301						
	g	0.762						
Np	s	6.821741	-0.133147	-0.133147	-0.133147	3.445732	0.675578	0.675578
		4.547828	0.631925	0.631925	0.631925	2.297155	-1.457351	-1.457351
		2.515128	1.000000	-1.119127	-1.119127	0.820301	1.000000	0.767844
		0.593735	1.000000	1.000000	1.002347	0.366318	1.000000	0.768379
		0.291083	1.000000	1.000000	0.379999	0.085724	1.000000	1.000000
		0.082683	1.000000	1.000000	1.000000	0.033821	1.000000	1.000000
		0.032691	1.000000	1.000000	1.000000			
	p	3.315850	0.163057	0.163057	0.163057	3.430082	0.150405	0.150405
		2.210567	-0.442359	-0.442359	-0.442359	2.286721	-0.394064	-0.394064

Table B.3: (continued).

An		(7s6p5d)/	[6s5p4d]	[5s4p4d]	[4s3p3d]	(6s5p4d)/	[5s4p3d]	[4s3p3d]
		0.831837	1.000000	0.335681	0.335681	0.622849	1.000000	0.675053
		0.446741	1.000000	1.000000	0.594773	0.281444	1.000000	0.456797
		0.223742	1.000000	1.000000	0.261454	0.112495	1.000000	1.000000
		0.098741	1.000000	1.000000	1.000000			
	d	1.172505	-0.166827	-0.166827	-0.166827	1.574459	-0.048339	-0.048339
		0.781670	0.295898	0.295898	0.295898	0.488698	0.418756	0.418756
		0.324572	1.000000	1.000000	0.442696	0.197216	1.000000	1.000000
		0.150000	1.000000	1.000000	1.000000	0.072854	1.000000	1.000000
		0.061817	1.000000	1.000000	1.000000			
	f	0.956						
		0.302						
	g	0.790						
Pu	s	7.352458	-0.116471	-0.116471	-0.116471	3.705929	0.629616	0.629616
		4.901638	0.549019	0.549019	0.549019	2.470619	-1.344135	-1.344135
		2.612308	1.000000	-1.039983	-1.039983	0.785316	1.000000	0.788786
		0.620947	1.000000	1.000000	0.988376	0.365895	1.000000	0.685647
		0.302882	1.000000	1.000000	0.384797	0.088023	1.000000	1.000000
		0.085489	1.000000	1.000000	1.000000	0.034539	1.000000	1.000000
		0.033617	1.000000	1.000000	1.000000			
	p	3.523566	0.151036	0.151036	0.151036	3.661456	0.141643	0.141643
		2.349044	-0.407907	-0.407907	-0.407907	2.440971	-0.368101	-0.368101
		0.808614	1.000000	0.380308	0.380308	0.636395	1.000000	0.677317
		0.439947	1.000000	1.000000	0.554985	0.286380	1.000000	0.443669
		0.223619	1.000000	1.000000	0.242489	0.115306	1.000000	1.000000
		0.098586	1.000000	1.000000	1.000000			
	d	1.227089	-0.157613	-0.157613	-0.157613	1.689265	-0.042046	-0.042046
		0.818059	0.291867	0.291867	0.291867	0.510712	0.416310	0.416310
		0.337965	1.000000	1.000000	0.440724	0.204181	1.000000	1.000000
		0.154987	1.000000	1.000000	1.000000	0.074681	1.000000	1.000000
		0.063342	1.000000	1.000000	1.000000			
	f	0.952						
		0.305						
	g	0.819						
Am	s	7.733388	-0.113694	-0.113694	-0.113694	3.928284	0.612700	0.612700
		5.155592	0.536989	0.536989	0.536989	2.618856	-1.299944	-1.299944
		2.739680	1.000000	-1.025253	-1.025253	0.785344	1.000000	0.815675
		0.650629	1.000000	1.000000	0.979339	0.370235	1.000000	0.634612
		0.315622	1.000000	1.000000	0.390165	0.090678	1.000000	1.000000
		0.088756	1.000000	1.000000	1.000000	0.035387	1.000000	1.000000

Table B.3: (continued).

An		(7s6p5d)/	[6s5p4d]	[5s4p4d]	[4s3p3d]	(6s5p4d)/	[5s4p3d]	[4s3p3d]
		0.034674	1.000000	1.000000	1.000000			
	p	3.727995	0.141694	0.141694	0.141694	3.849157	0.134723	0.134723
		2.485330	-0.380163	-0.380163	-0.380163	2.566105	-0.350564	-0.350564
		0.785834	1.000000	0.440094	0.440094	0.657137	1.000000	0.673531
		0.423760	1.000000	1.000000	0.514598	0.294044	1.000000	0.441799
		0.218808	1.000000	1.000000	0.213989	0.117875	1.000000	1.000000
		0.096512	1.000000	1.000000	1.000000			
	d	1.295391	-0.150139	-0.150139	-0.150139	1.807675	-0.038149	-0.038149
		0.863594	0.281570	0.281570	0.281570	0.532299	0.413741	0.413741
		0.355154	1.000000	1.000000	0.437960	0.210843	1.000000	1.000000
		0.161330	1.000000	1.000000	1.000000	0.076355	1.000000	1.000000
		0.065160	1.000000	1.000000	1.000000			
	f	0.952						
		0.310						
	g	0.847						

Table B.4: Exponents and contraction coefficients of the (7s6p5d), (7s6p5d)/[6s5p4d] {211111/21111/2111}, (7s6p5d)/[5s4p4d] {31111/3111/2111}, and (7s6p5d)/[4s3p3d] {3211/321/311} as well as of the (6s5p4d), (6s5p4d)/[5s4p3d] {21111/2111/211}, and (6s5p4d)/[4s3p3d] {2211/221/211} GTO valence basis sets for the hexavalent ($5f^{n-3}$, n=3–6 for U–Am) 5f-in-core PPs for actinides [30]. Additionally, the two f and one g polarization functions are given.

An		(7s6p5d)/	[6s5p4d]	[5s4p4d]	[4s3p3d]	(6s5p4d)/	[5s4p3d]	[4s3p3d]
U	s	5.777619	-0.224778	-0.224778	-0.224778	2.965398	0.942263	0.942263
		3.851500	1.181527	1.181527	1.181527	1.976431	-2.250561	-2.250561
		2.499039	1.000000	-1.653779	-1.653779	1.087315	1.000000	1.074213
		0.597383	1.000000	1.000000	1.050747	0.407198	1.000000	0.963501
		0.303537	1.000000	1.000000	0.390081	0.090236	1.000000	1.000000
		0.085200	1.000000	1.000000	1.000000	0.036147	1.000000	1.000000
		0.033918	1.000000	1.000000	1.000000			
	p	5.053602	-0.019953	-0.019953	-0.019953	3.215541	0.173929	0.173929
		3.369068	0.225455	0.225455	0.225455	2.143694	-0.450876	-0.450876
		2.246046	1.000000	-0.471844	-0.471844	0.628030	1.000000	0.683015
		0.614521	1.000000	1.000000	0.691787	0.297698	1.000000	0.455291
		0.292693	1.000000	1.000000	0.437970	0.128416	1.000000	1.000000
		0.127542	1.000000	1.000000	1.000000			
	d	1.090623	-0.170429	-0.170429	-0.170429	1.347682	-0.059524	-0.059524

Table B.4: (continued).

An		(7s6p5d)/	[6s5p4d]	[5s4p4d]	[4s3p3d]	(6s5p4d)/	[5s4p3d]	[4s3p3d]
		0.727082	0.327830	0.327830	0.327830	0.510652	0.411309	0.411309
		0.324934	1.000000	1.000000	0.426911	0.219840	1.000000	1.000000
		0.160567	1.000000	1.000000	1.000000	0.083802	1.000000	1.000000
		0.067945	1.000000	1.000000	1.000000			
	f	1.199						
		0.356						
	g	0.766						
Np	s	6.270359	-0.192963	-0.192963	-0.192963	3.171367	0.885207	0.885207
		4.179579	0.968049	0.968049	0.968049	2.114245	-2.070958	-2.070958
		2.568653	1.000000	-1.461026	-1.461026	1.104517	1.000000	0.967041
		0.627355	1.000000	1.000000	1.031224	0.422014	1.000000	0.951448
		0.317619	1.000000	1.000000	0.401521	0.093268	1.000000	1.000000
		0.088178	1.000000	1.000000	1.000000	0.037118	1.000000	1.000000
		0.034900	1.000000	1.000000	1.000000			
	p	5.305896	-0.017012	-0.017012	-0.017012	3.391119	0.165078	0.165078
		3.537264	0.209254	0.209254	0.209254	2.260746	-0.428221	-0.428221
		2.358176	1.000000	-0.446389	-0.446389	0.647970	1.000000	0.678711
		0.635597	1.000000	1.000000	0.687027	0.305647	1.000000	0.451108
		0.300929	1.000000	1.000000	0.435738	0.131542	1.000000	1.000000
		0.130713	1.000000	1.000000	1.000000			
	d	1.127203	-0.165108	-0.165108	-0.165108	1.436375	-0.052455	-0.052455
		0.751468	0.337442	0.337442	0.337442	0.532570	0.409933	0.409933
		0.331220	1.000000	1.000000	0.429555	0.227582	1.000000	1.000000
		0.162417	1.000000	1.000000	1.000000	0.086071	1.000000	1.000000
		0.068661	1.000000	1.000000	1.000000			
	f	1.066						
		0.329						
	g	0.795						
Pu	s	6.824056	-0.165622	-0.165622	-0.165622	3.424235	0.809925	0.809925
		4.549333	0.806186	0.806186	0.806186	2.282824	-1.842262	-1.842262
		2.665742	1.000000	-1.305994	-1.305994	1.089376	1.000000	0.845177
		0.651837	1.000000	1.000000	1.016481	0.433681	1.000000	0.924309
		0.328159	1.000000	1.000000	0.399895	0.096549	1.000000	1.000000
		0.091144	1.000000	1.000000	1.000000	0.038120	1.000000	1.000000
		0.035874	1.000000	1.000000	1.000000			
	p	5.563405	-0.012302	-0.012302	-0.012302	3.591427	0.155230	0.155230
		3.708937	0.187517	0.187517	0.187517	2.394284	-0.402625	-0.402625
		2.472624	1.000000	-0.416448	-0.416448	0.667857	1.000000	0.673726
		0.658940	1.000000	1.000000	0.679455	0.313485	1.000000	0.446390

Table B.4: (continued).

An		(7s6p5d)/	[6s5p4d]	[5s4p4d]	[4s3p3d]	(6s5p4d)/	[5s4p3d]	[4s3p3d]
		0.310025	1.000000	1.000000	0.436073	0.134646	1.000000	1.000000
		0.133880	1.000000	1.000000	1.000000			
	d	1.177699	-0.155539	-0.155539	-0.155539	1.526316	-0.045974	-0.045974
		0.785133	0.331259	0.331259	0.331259	0.554588	0.408291	0.408291
		0.345487	1.000000	1.000000	0.426459	0.235205	1.000000	1.000000
		0.168585	1.000000	1.000000	1.000000	0.088229	1.000000	1.000000
		0.070697	1.000000	1.000000	1.000000			
	f	1.032						
		0.327						
	g	0.823						
Am	s	7.328600	-0.149786	-0.149786	-0.149786	3.676233	0.749578	0.749578
		4.885732	0.712613	0.712613	0.712613	2.450822	-1.659002	-1.659002
		2.762048	1.000000	-1.217373	-1.217373	1.036368	1.000000	0.784016
		0.679956	1.000000	1.000000	1.004711	0.439234	1.000000	0.867115
		0.340677	1.000000	1.000000	0.403490	0.099186	1.000000	1.000000
		0.094277	1.000000	1.000000	1.000000	0.038906	1.000000	1.000000
		0.036909	1.000000	1.000000	1.000000			
	p	5.806696	-0.008625	-0.008625	-0.008625	3.779297	0.147744	0.147744
		3.871130	0.170603	0.170603	0.170603	2.519531	-0.382953	-0.382953
		2.580754	1.000000	-0.392918	-0.392918	0.688466	1.000000	0.670314
		0.682076	1.000000	1.000000	0.674121	0.321509	1.000000	0.442971
		0.319005	1.000000	1.000000	0.436164	0.137679	1.000000	1.000000
		0.136981	1.000000	1.000000	1.000000			
	d	1.214926	-0.149320	-0.149320	-0.149320	1.591694	-0.041863	-0.041863
		0.809950	0.330653	0.330653	0.330653	0.574192	0.407967	0.407967
		0.356055	1.000000	1.000000	0.423872	0.241895	1.000000	1.000000
		0.173127	1.000000	1.000000	1.000000	0.090084	1.000000	1.000000
		0.072173	1.000000	1.000000	1.000000			
	f	1.014						
		0.327						
	g	0.849						

Table B.5: Energy differences (in eV) between finite-difference [55] and divalent ($5f^{n+1}$, n=5–13 for Pu–No; $Q = 10$) PP HF calculations [57] using (4s4p) and (5s5p) as well as (6s5p4d) and (7s6p5d) valence basis sets to calculate the configurations An^{2+}: $6s^26p^6$ and An: $6s^26p^66d^17s^1$, respectively [29]. For the tetravalent ($5f^{n-1}$, n=1–9 for Th–Cf; $Q = 12$) PPs the energy differences between finite-difference and PP HF calculations using (4s4p3d) and (5s5p4d) as well as (6s5p4d) and (7s6p5d) valence basis sets to calculate the configurations An^{3+}: $6s^26p^66d^1$ and An: $6s^26p^66d^27s^2$ are given, respectively [29].

| | $Q = 10$ | | | | $Q = 12$ | | | |
| | An^{2+}: $6s^26p^6$ | | An: $6s^26p^66d^17s^1$ | | An^{3+}: $6s^26p^66d^1$ | | An: $6s^26p^66d^27s^2$ | |
An	(4s4p)	(5s5p)	(6s5p4d)	(7s6p5d)	(4s4p3d)	(5s5p4d)	(6s5p4d)	(7s6p5d)
Th					0.088	0.030	0.084	0.025
Pa					0.091	0.027	0.089	0.025
U					0.088	0.028	0.085	0.028
Np					0.088	0.031	0.083	0.034
Pu	0.059	0.007	0.066	0.011	0.098	0.034	0.092	0.039
Am	0.060	0.008	0.070	0.012	0.107	0.039	0.101	0.045
Cm	0.062	0.010	0.074	0.014	0.115	0.047	0.109	0.055
Bk	0.069	0.012	0.084	0.017	0.127	0.058	0.119	0.068
Cf	0.078	0.014	0.096	0.019	0.140	0.065	0.132	0.077
Es	0.079	0.019	0.102	0.025				
Fm	0.087	0.022	0.115	0.029				
Md	0.095	0.025	0.130	0.033				
No	0.105	0.026	0.145	0.035				

Table B.6: Energy differences (in eV) between finite-difference [55] and pentavalent ($5f^{n-2}$, n=2–6 for Pa–Am; $Q = 13$) PP HF calculations [57] using (4s4p3d) and (5s5p4d) as well as (6s5p4d) and (7s6p5d) valence basis sets to calculate the configurations An^{4+}: $6s^26p^66d^1$ and An: $6s^26p^66d^37s^2$, respectively [30]. For the hexavalent ($5f^{n-3}$, n=3–6 for U–Am; $Q = 14$) PPs the energy differences between finite-difference and PP HF calculations using (4s4p3d) and (5s5p4d) as well as (6s5p4d) and (7s6p5d) valence basis sets to calculate the configurations An^{5+}: $6s^26p^66d^1$ and An: $6s^26p^66d^47s^2$ are given, respectively [30].

| | $Q = 13$ | | | | $Q = 14$ | | | |
| | An^{4+}: $6s^26p^66d^1$ | | An: $6s^26p^66d^37s^2$ | | An^{5+}: $6s^26p^66d^1$ | | An: $6s^26p^66d^47s^2$ | |
An	(4s4p3d)	(5s5p4d)	(6s5p4d)	(7s6p5d)	(4s4p3d)	(5s5p4d)	(6s5p4d)	(7s6p5d)
Pa	0.089	0.012	0.103	0.021				
U	0.087	0.011	0.100	0.020	0.120	0.021	0.153	0.040
Np	0.086	0.011	0.097	0.019	0.113	0.023	0.145	0.042
Pu	0.079	0.011	0.090	0.019	0.106	0.025	0.139	0.045
Am	0.080	0.011	0.091	0.020	0.104	0.028	0.139	0.049

Table B.11: Exponents and contraction coefficients of the (7s6p5d), (7s6p5d)/[6s5p4d] {211111/21111/2111}, (7s6p5d)/[5s4p4d] {31111/2211/2111}, and (7s6p5d)/[4s3p3d] {3211/411/311} as well as of the (6s5p4d), (6s5p4d)/[5s4p3d] {21111/2111/211}, and (6s5p4d)/[4s3p3d] {3111/311/211} GTO valence basis sets[a] for the tetravalent ($4f^{n-1}$, n=1–3, 8, 9 for Ce–Nd, Tb, Dy) 4f-in-core PPs for lanthanides [28]. Additionally, the two f and one g polarization functions are given.

Ln		(7s6p5d)/	[6s5p4d]	[5s4p4d]	[4s3p3d]	(6s5p4d)/	[5s4p3d]	[4s3p3d]
Ce	s	6.498173	-0.164521	-0.164521	-0.164521	3.333203	0.640046	0.640046

Table B.11: (continued).

Ln		(7s6p5d)/	[6s5p4d]	[5s4p4d]	[4s3p3d]	(6s5p4d)/	[5s4p3d]	[4s3p3d]
		4.332116	0.712041	0.712041	0.712041	2.222136	-1.317178	-1.317178
		2.454114	1.000000	-1.120622	-1.120622	0.658203	1.000000	0.790993
		0.521019	1.000000	1.000000	0.982403	0.301946	1.000000	1.000000
		0.248300	1.000000	1.000000	0.364449	0.062055	1.000000	1.000000
		0.064112	1.000000	1.000000	1.000000	0.025345	1.000000	1.000000
		0.025705	1.000000	1.000000	1.000000			
	p	3.656443	0.155547	0.155547	0.155547	3.783090	0.145862	0.145862
		2.437629	-0.411357	-0.411357	-0.411357	2.522060	-0.373683	-0.373683
		0.804966	1.000000	0.369656	0.369656	0.623777	1.000000	0.677334
		0.420218	1.000000	0.581143	0.581143	0.266554	1.000000	1.000000
		0.204082	1.000000	1.000000	1.000000	0.092153	1.000000	1.000000
		0.083900	1.000000	1.000000	1.000000			
	d	1.454210	-0.209427	-0.209427	-0.209427	1.937939	-0.067371	-0.067371
		0.969473	0.297463	0.297463	0.297463	0.527082	0.453919	0.453919
		0.358359	1.000000	1.000000	0.485714	0.190426	1.000000	1.000000
		0.150000	1.000000	1.000000	1.000000	0.066784	1.000000	1.000000
		0.060100	1.000000	1.000000	1.000000			
	f	0.992						
		0.324						
	g	0.835						
Pr	s	6.764679	-0.156672	-0.156672	-0.156672	3.498540	0.622914	0.622914
		4.509786	0.696973	0.696973	0.696973	2.332360	-1.286578	-1.286578
		2.576083	1.000000	-1.103512	-1.103512	0.683341	1.000000	0.784805
		0.542333	1.000000	1.000000	0.974049	0.312246	1.000000	1.000000
		0.257175	1.000000	1.000000	0.366060	0.063938	1.000000	1.000000
		0.066033	1.000000	1.000000	1.000000	0.025982	1.000000	1.000000
		0.026344	1.000000	1.000000	1.000000			
	p	3.839265	0.148110	0.148110	0.148110	4.000062	0.139431	0.139431
		2.559510	-0.394349	-0.394349	-0.394349	2.666708	-0.357004	-0.357004
		0.826273	1.000000	0.378055	0.378055	0.642891	1.000000	0.677207
		0.429633	1.000000	0.572263	0.572263	0.273219	1.000000	1.000000
		0.208455	1.000000	1.000000	1.000000	0.094304	1.000000	1.000000
		0.085533	1.000000	1.000000	1.000000			
	d	1.495839	-0.235458	-0.235458	-0.235458	2.088603	-0.061917	-0.061917
		1.048004	0.325735	0.325735	0.325735	0.549009	0.450503	0.450503
		0.369529	1.000000	1.000000	0.493464	0.196381	1.000000	1.000000
		0.151566	1.000000	1.000000	1.000000	0.068210	1.000000	1.000000
		0.060333	1.000000	1.000000	1.000000			
	f	0.995						

Table B.11: (continued).

Ln		(7s6p5d)/	[6s5p4d]	[5s4p4d]	[4s3p3d]	(6s5p4d)/	[5s4p3d]	[4s3p3d]
		0.327						
	g	0.866						
Nd	s	7.125040	-0.139986	-0.139986	-0.139986	3.670677	0.598862	0.598862
		4.750027	0.637428	0.637428	0.637428	2.447118	-1.247317	-1.247317
		2.687669	1.000000	-1.045597	-1.045597	0.717401	1.000000	0.760311
		0.561486	1.000000	1.000000	0.964707	0.325254	1.000000	1.000000
		0.264971	1.000000	1.000000	0.364776	0.065757	1.000000	1.000000
		0.067869	1.000000	1.000000	1.000000	0.026615	1.000000	1.000000
		0.026954	1.000000	1.000000	1.000000			
	p	4.039866	0.139883	0.139883	0.139883	4.222577	0.132128	0.132128
		2.693244	-0.375612	-0.375612	-0.375612	2.815051	-0.340008	-0.340008
		0.850954	1.000000	0.382121	0.382121	0.665613	1.000000	0.673530
		0.440767	1.000000	0.565268	0.565268	0.281142	1.000000	1.000000
		0.213406	1.000000	1.000000	1.000000	0.096418	1.000000	1.000000
		0.087318	1.000000	1.000000	1.000000			
	d	1.685274	-0.153534	-0.153534	-0.153534	2.250932	-0.056678	-0.056678
		1.003460	0.259724	0.259724	0.259724	0.571488	0.446659	0.446659
		0.375777	1.000000	1.000000	0.486316	0.202338	1.000000	1.000000
		0.154095	1.000000	1.000000	1.000000	0.069586	1.000000	1.000000
		0.061077	1.000000	1.000000	1.000000			
	f	1.007						
		0.332						
	g	0.899						
Tb	s	8.864395	-0.091411	-0.091411	-0.091411	4.733798	0.497666	0.497666
		5.909597	0.468468	0.468468	0.468468	3.155865	-1.036011	-1.036011
		3.249746	1.000000	-0.881094	-0.881094	0.742729	1.000000	0.837221
		0.677911	1.000000	1.000000	0.918820	0.343473	1.000000	1.000000
		0.313284	1.000000	1.000000	0.383016	0.077122	1.000000	1.000000
		0.078044	1.000000	1.000000	1.000000	0.030207	1.000000	1.000000
		0.030329	1.000000	1.000000	1.000000			
	p	5.135357	0.109188	0.109188	0.109188	5.350305	0.105396	0.105396
		3.417168	-0.299934	-0.299934	-0.299934	3.566870	-0.276891	-0.276891
		0.953081	1.000000	0.422822	0.422822	0.787431	1.000000	0.656596
		0.483900	1.000000	0.521816	0.521816	0.322548	1.000000	1.000000
		0.234632	1.000000	1.000000	1.000000	0.106344	1.000000	1.000000
		0.094850	1.000000	1.000000	1.000000			
	d	2.486442	-0.069396	-0.069396	-0.069396	3.116672	-0.038069	-0.038069
		0.999118	0.235607	0.235607	0.235607	0.685808	0.425550	0.425550
		0.403325	1.000000	1.000000	0.457161	0.230483	1.000000	1.000000

Table B.11: (continued).

Ln		(7s6p5d)/	[6s5p4d]	[5s4p4d]	[4s3p3d]	(6s5p4d)/	[5s4p3d]	[4s3p3d]
		0.164460	1.000000	1.000000	1.000000	0.075422	1.000000	1.000000
		0.063727	1.000000	1.000000	1.000000			
	f	1.092						
		0.356						
	g	1.062						
Dy	s	9.036326	-0.086301	-0.086301	-0.086301	4.946819	0.485099	0.485099
		6.024218	0.475836	0.475836	0.475836	3.297879	-1.011025	-1.011025
		3.405647	1.000000	-0.885180	-0.885180	0.758794	1.000000	0.844454
		0.700604	1.000000	1.000000	0.914258	0.349639	1.000000	1.000000
		0.322309	1.000000	1.000000	0.382143	0.079463	1.000000	1.000000
		0.080304	1.000000	1.000000	1.000000	0.030956	1.000000	1.000000
		0.031066	1.000000	1.000000	1.000000			
	p	5.376580	0.102722	0.102722	0.102722	5.580222	0.101635	0.101635
		3.555459	-0.287060	-0.287060	-0.287060	3.720148	-0.267577	-0.267577
		0.974988	1.000000	0.433087	0.433087	0.813232	1.000000	0.653770
		0.489507	1.000000	0.517024	0.517024	0.331161	1.000000	1.000000
		0.236874	1.000000	1.000000	1.000000	0.108282	1.000000	1.000000
		0.096092	1.000000	1.000000	1.000000			
	d	2.628502	-0.064009	-0.064009	-0.064009	3.296408	-0.035489	-0.035489
		1.020405	0.235625	0.235625	0.235625	0.708915	0.421208	0.421208
		0.410965	1.000000	1.000000	0.453028	0.235803	1.000000	1.000000
		0.166876	1.000000	1.000000	1.000000	0.076403	1.000000	1.000000
		0.064225	1.000000	1.000000	1.000000			
	f	1.113						
		0.361						
	g	1.096						

[a] The valence basis sets were optimized by M. Hülsen.

Table B.7: Energy differences (in eV) between finite-difference [55] and divalent ($5f^{n+1}$, n=5–13 for Pu–No; $Q = 10$) PP HF calculations [52] using (6s5p4d)/[4s3p3d], (6s5p4d)/[5s4p3d], (7s6p5d)/[4s3p3d], (7s6p5d)/[5s4p4d], and (7s6p5d)/[6s5p4d] valence basis sets to calculate the configuration An: $6s^2 6p^6 6d^1 7s^1$ [29].

An	(6s5p4d)/		(7s6p5d)/		
	[4s3p3d]	[5s4p3d]	[4s3p3d]	[5s4p4d]	[6s5p4d]
Pu	0.081	0.071	0.024	0.015	0.011
Am	0.085	0.074	0.026	0.016	0.012
Cm	0.090	0.078	0.029	0.019	0.015
Bk	0.102	0.088	0.034	0.021	0.017
Cf	0.118	0.100	0.040	0.024	0.020
Es	0.122	0.106	0.044	0.029	0.025
Fm	0.138	0.119	0.051	0.033	0.029
Md	0.155	0.134	0.058	0.038	0.034
No	0.177	0.150	0.066	0.041	0.035

Table B.8: Energy differences (in eV) between finite-difference [55] and tetravalent ($5f^{n-1}$, n=1–9 for Th–Cf; $Q = 12$) PP HF calculations [52] using (6s5p4d)/[4s3p3d], (6s5p4d)/[5s4p3d], (7s6p5d)/[4s3p3d], (7s6p5d)/[5s4p4d], and (7s6p5d)/[6s5p4d] valence basis sets to calculate the configuration An: $6s^2 6p^6 6d^2 7s^2$ [29].

An	(6s5p4d)/		(7s6p5d)/		
	[4s3p3d]	[5s4p3d]	[4s3p3d]	[5s4p4d]	[6s5p4d]
Th	0.122	0.117	0.057	0.057	0.026
Pa	0.127	0.120	0.057	0.056	0.026
U	0.118	0.113	0.058	0.056	0.029
Np	0.112	0.108	0.061	0.060	0.035
Pu	0.124	0.117	0.068	0.065	0.040
Am	0.135	0.126	0.077	0.071	0.046
Cm	0.143	0.132	0.087	0.080	0.056
Bk	0.153	0.142	0.100	0.093	0.069
Cf	0.172	0.157	0.115	0.104	0.079

Table B.9: Energy differences (in eV) between finite-difference [55] and pentavalent ($5f^{n-2}$, n=2–6 for Pa–Am; $Q = 13$) PP HF calculations [52] using (6s5p4d)/[4s3p3d], (6s5p4d)/[5s4p3d], (7s6p5d)/[4s3p3d], (7s6p5d)/[5s4p4d], and (7s6p5d)/[6s5p4d] valence basis sets to calculate the configuration An: $6s^2 6p^6 6d^3 7s^2$ [30].

An	(6s5p4d)/		(7s6p5d)/		
	[4s3p3d]	[5s4p3d]	[4s3p3d]	[5s4p4d]	[6s5p4d]
Pa	0.164	0.159	0.075	0.075	0.023
U	0.157	0.152	0.071	0.071	0.022
Np	0.157	0.148	0.068	0.067	0.021
Pu	0.139	0.134	0.064	0.063	0.021
Am	0.137	0.132	0.063	0.062	0.021

Table B.10: Energy differences (in eV) between finite-difference [55] and hexavalent ($5f^{n-3}$, n=3–6 for U–Am; $Q = 14$) PP HF calculations [52] using (6s5p4d)/[4s3p3d], (6s5p4d)/[5s4p3d], (7s6p5d)/[4s3p3d], (7s6p5d)/[5s4p4d], and (7s6p5d)/[6s5p4d] valence basis sets to calculate the configuration An: $6s^2 6p^6 6d^4 7s^2$ [30].

An	(6s5p4d)/		(7s6p5d)/		
	[4s3p3d]	[5s4p3d]	[4s3p3d]	[5s4p4d]	[6s5p4d]
U	0.324	0.249	0.125	0.124	0.044
Np	0.290	0.236	0.121	0.120	0.046
Pu	0.251	0.220	0.115	0.115	0.048
Am	0.228	0.212	0.114	0.114	0.051

Table B.12: Energy differences (in eV) between finite-difference [55] and tetravalent ($4f^{n-1}$, n=1–3, 8, 9 for Ce–Nd, Tb, Dy) PP HF calculations [57] using (4s4p3d) and (5s5p4d) as well as (6s5p4d) and (7s6p5d) valence basis sets to calculate the configurations Ln^{3+}: $5s^2 5p^6 5d^1$ and Ln: $5s^2 5p^6 5d^2 6s^2$, respectively [28].

Ln	Ln^{3+}: $5s^2 5p^6 5d^1$		Ln: $5s^2 5p^6 5d^2 6s^2$	
	(4s4p3d)	(5s5p4d)	(6s5p4d)	(7s6p5d)
Ce	0.149	0.021	0.146	0.026
Pr	0.144	0.019	0.138	0.024
Nd	0.135	0.018	0.126	0.022
Tb	0.133	0.017	0.110	0.021
Dy	0.138	0.019	0.112	0.022

Table B.13: Energy differences (in eV) between finite-difference [55] and tetravalent ($4f^{n-1}$, n=1–3, 8, 9 for Ce–Nd, Tb, Dy) PP HF calculations [52] using (6s5p4d)/[4s3p3d], (6s5p4d)/[5s4p3d], (7s6p5d)/[4s3p3d], (7s6p5d)/[5s4p4d], and (7s6p5d)/[6s5p4d] valence basis sets to calculate the configuration Ln: $5s^2 5p^6 5d^2 6s^2$ [28].

Ln	(6s5p4d)/		(7s6p5d)/		
	[4s3p3d]	[5s4p3d]	[4s3p3d]	[5s4p4d]	[6s5p4d]
Ce	0.188	0.167	0.064	0.047	0.027
Pr	0.181	0.158	0.061	0.043	0.025
Nd	0.168	0.144	0.056	0.039	0.023
Tb	0.193	0.123	0.053	0.034	0.022
Dy	0.203	0.124	0.054	0.035	0.023

Table B.14: Exponents and contraction coefficients (coeff.) of the (14s13p10d8f6g)/[6s6p5d4f3g] generalized contracted ANO valence basis set[a] of the 5f-in-valence MCDHF/DCB PP for uranium [16].

	Exponents	Coeff.	Coeff.	Coeff.	Coeff.	Coeff.	Coeff.
s	29520.8340	0.000044	-0.000033	0.000014	-0.000040	-0.000052	-0.000106
	4449.8874	0.000320	-0.000252	0.000103	-0.000298	-0.000339	-0.000790
	1018.7754	0.001395	-0.001015	0.000411	-0.001259	-0.001749	-0.003263

Table B.14: (continued).

	Exponents	Coeff.	Coeff.	Coeff.	Coeff.	Coeff.	Coeff.
	289.5348	0.002814	-0.002415	0.001004	-0.002775	-0.002574	-0.007528
	46.9990	0.096435	-0.039561	0.014267	-0.050598	-0.114002	-0.096221
	33.5707	-0.487967	0.218831	-0.081043	0.253466	0.460601	0.485331
	23.9791	0.807508	-0.397990	0.150715	-0.440238	-0.667382	-0.882533
	10.1441	-1.127504	0.664738	-0.259766	0.745253	0.973604	1.727216
	2.7658	1.060527	-1.013418	0.420775	-1.600118	-2.866306	-7.334892
	1.4570	0.400292	-0.393365	0.189368	0.109425	2.637145	11.247552
	0.6151	0.016450	0.904344	-0.540005	2.867628	0.781971	-9.107755
	0.2780	-0.000219	0.518022	-0.412368	-2.240568	-2.330988	4.807540
	0.0570	-0.000042	0.073065	0.784983	-0.564799	2.737967	-1.837764
	0.0233	-0.000176	0.029494	0.424189	1.005222	-1.958184	0.966838
p	499.7488	0.000219	-0.000127	0.000103	-0.000219	-0.000595	0.000338
	114.0192	0.001190	-0.000643	0.000473	-0.000353	0.001246	0.002000
	21.9088	-0.042782	0.018120	-0.011387	-0.018092	-0.195716	-0.062289
	15.6492	0.324353	-0.166453	0.121930	-0.055835	0.559357	0.603816
	11.1780	-0.511529	0.306811	-0.234776	0.157684	-0.775626	-1.392522
	7.9843	-0.071299	-0.014845	0.011640	0.140305	0.968111	0.608885
	3.1325	0.704008	-0.476924	0.408928	-0.907514	-2.680683	2.933431
	1.6077	0.465061	-0.276612	0.240978	0.013107	2.940086	-5.469685
	0.7061	0.054229	0.480908	-0.888499	1.811441	-0.333255	5.883058
	0.3229	-0.001819	0.615543	-0.081066	-1.372113	-2.202343	-4.876290
	0.1329	0.002302	0.158324	0.350416	-1.074410	3.526594	3.141966
	0.0800	-0.001129	-0.005358	0.613368	1.296881	-1.910003	-0.903616
	0.0271	0.000143	0.003348	0.144497	0.249768	-0.256716	-0.504585
d	258.1524	0.000136	-0.000042	-0.000027	-0.000293	-0.000057	
	75.1703	0.001076	-0.000450	-0.001045	0.003467	-0.001967	
	20.7869	0.012136	-0.003939	-0.004760	-0.009267	-0.010974	
	7.6295	-0.229921	0.084412	0.128096	-0.010746	0.427668	
	5.5167	0.278231	-0.113456	-0.162964	-0.060234	-0.708332	
	2.6058	0.551975	-0.213270	-0.463388	1.266130	-1.037124	
	1.2781	0.377844	-0.100722	-0.054235	-1.508105	2.617986	
	0.5617	0.074069	0.321783	1.072575	0.062980	-2.545826	
	0.2135	0.000082	0.561639	-0.281443	1.054738	1.686553	
	0.0719	0.000484	0.344991	-0.608317	-0.898798	-0.787351	
f	59.3666	0.001424	-0.001690	-0.002261	-0.000347		
	20.3890	0.011120	-0.011132	-0.018893	0.030982		
	8.1761	0.041970	-0.050308	-0.067853	-0.008213		
	3.5111	0.207732	-0.221735	-0.414443	0.721303		
	1.6789	0.370755	-0.378340	-0.431246	-0.469218		

Table B.14: (continued).

	Exponents	Coeff.	Coeff.	Coeff.	Coeff.	Coeff.	Coeff.
	0.7604	0.373775	-0.013232	1.002423	-0.747053		
	0.3170	0.252996	0.525098	-0.036904	1.376481		
	0.1149	0.097955	0.473370	-0.612297	-0.944182		
g	59.3666	0.001014	0.003561	-0.000733			
	20.3890	0.011848	0.028234	-0.075550			
	8.1761	0.039302	0.137274	-0.064958			
	3.5111	0.124196	0.304284	-1.260170			
	1.6789	0.233870	0.787343	1.661908			
	0.7604	0.684875	-0.903712	-0.787501			

[a] The valence basis set was optimized by X. Cao.

Appendix C

Test Calculations and Applications

Table C.1: First and second IPs (in eV) for the actinides, where the 5f occupation stays constant, from 5f-in-core LPP CCSD(T) calculations [52] with and without using CPPs in comparison to experimental [59–61] and SPP multi-reference ACPF calculations without SO coupling at the basis set limit [21], respectively. In the LPP calculations (7s6p5d2f1g) basis sets were applied. Additionally, the m.a.e. (in eV) and m.r.e. are given with respect to experimental and SPP data for IP_1 and IP_2, respectively.

An	IP_1			IP_2		
	LPP	CPP[a]	Exp.	LPP	CPP[a]	SPP
Ac	4.99	4.94	5.17±0.12	11.68	11.71	11.78±0.19[b]
Th	6.11	6.15	6.31			
Pa	5.13	5.09	5.90±0.12	12.51	12.67	12.07[c]
U	5.13	5.09	6.19			
Np	6.19	6.24	6.27			
Pu	5.77	5.89	6.03	11.28	11.54	11.55
Am	5.84	5.97	5.97	11.43	11.69	11.71
Cm	4.92	4.94	5.99			
Bk	5.98	6.10	6.20	11.72	11.98	11.97
Cf	6.05	6.17	6.28	11.86	12.11	12.04
Es	6.12	6.24	6.37	12.02	12.25	12.20
Fm	6.19	6.30	6.50±0.07	12.16	12.39	12.38
Md	6.26	6.37	6.58	12.31	12.53	12.46
No	6.33	6.44	6.65	12.45	12.64	12.58
Lr	3.42	3.43	3.97[d]	14.24	14.35	14.24
m.a.e.	0.40	0.33		0.20	0.10	
m.r.e.	6.8%	5.8%		1.7%	0.8%	

[a]LPP calculations using CPPs.
[b]Experimental value.
[c]SPP calculation using the standard basis set (14s13p10d8f6g)/[6s6p5d4f3g].
[d]SPP CCSD(T) value.

Table C.2: Experimentally observed ground states [71] and AE WB energy differences (in eV) between these ground states and the lowest valence substate corresponding to di- ($5f^{n+1}$, n=5–13 for Pu–No) [29], tri- ($5f^n$, n=0–14 for Ac–Lr) [54], and tetravalent ($5f^{n-1}$, n=1–9 for Th–Cf) [29] 5f-in-core LPPs, respectively.

	Ground State		$5f^{n+1}6s^26p^67s^2$		$5f^n6s^26p^66d^17s^2$		$5f^{n-1}6s^26p^66d^27s^2$	
An	Configuration	Term	Term	ΔE^a	Term^b	ΔE^c	Term	ΔE^d
Ac	$5f^06d^17s^2$	^2D			^2D	0.00		
Th	$5f^06d^27s^2$	^3F			^3H	1.31	^3F	0.00
Pa	$5f^26d^17s^2$	^4K			^4K	0.00	^4I	0.06
U	$5f^36d^17s^2$	^5L			^5L	0.00	^5L	1.08
Np	$5f^46d^17s^2$	^6L			^6L	0.00	^6M	1.49
Pu	$5f^67s^2$	^7F	^7F	0.00	^7K	-0.32	^7M	1.74
Am	$5f^77s^2$	^8S	^8S	0.00	^8H	1.54	^8L	5.12
Cm	$5f^76d^17s^2$	^9D	^7F	2.68	^9D	0.00	^9I	5.37
Bk	$5f^97s^2$	^6H	^6H	0.00	^8G		^{10}F	-0.04
Cf	$5f^{10}7s^2$	^5I	^5I	0.00	^7K	1.19	^9I	3.93
Es	$5f^{11}7s^2$	^4I	^4I	0.00	^6I			
Fm	$5f^{12}7s^2$	^3H	^3H	0.00	^5L	1.70		
Md	$5f^{13}7s^2$	^2F	^2F	0.00	^4K	3.08		
No	$5f^{14}7s^2$	^1S	^1S	0.00	^3H	5.08		
Lr	$5f^{14}7s^27p^1$	^2P			^2D	0.02		

$^a\Delta E = E(5f^{n+1}6s^26p^67s^2) - E(\text{ground state})$.
^{b}For Bk and Es no AE WB energy corrections could be calculated, because there is more than one possibility to couple $5f^n$ and $6d^1$ to obtain the desired LS-states ^8G and ^6I, respectively.
$^c\Delta E = E(5f^n6s^26p^66d^17s^2) - E(\text{ground state})$.
$^d\Delta E = E(5f^{n-1}6s^26p^66d^27s^2) - E(\text{ground state})$.

Table C.3: Experimentally observed ground states [71] and AE WB energy differences (in eV) between these ground states and the lowest valence substate corresponding to pentavalent 5f-in-core LPPs ($5f^{n-2}$, n=2–6 for Pa–Am) and the hexavalent LPP for uranium ($5f^{n-3}$, n=3), respectively [30].

	Ground State		$5f^{n-2}6s^26p^66d^37s^2$		$5f^{n-3}6s^26p^66d^47s^2$	
An	Configuration	Term	Term	ΔE^a	Term	ΔE^b
Pa	$5f^26d^17s^2$	^4K	^4F	2.39		
U	$5f^36d^17s^2$	^5L	^5I	5.29	^5D	11.86
Np	$5f^46d^17s^2$	^6L	^6L	8.21		
Pu	$5f^67s^2$	^7F	^7M	11.44		
Am	$5f^77s^2$	^8S	^8M	15.98		

$^a\Delta E = E(5f^{n-2}6s^26p^66d^37s^2) - E(\text{ground state})$.
$^b\Delta E = E(5f^{n-3}6s^26p^66d^47s^2) - E(\text{ground state})$.

Table C.4: Experimentally observed ground states [71] and AE DHF energy differences (in eV) between these ground states and the lowest valence substate corresponding to di- ($5f^{n+1}$, n=5–13 for Pu–No) [29], tri- ($5f^n$, n=0–14 for Ac–Lr) [54], and tetravalent ($5f^{n-1}$, n=1–9 for Th–Cf) [29] 5f-in-core LPPs, respectively.

	Ground State		$5f^{n+1}6s^26p^67s^2$		$5f^n6s^26p^66d^17s^2$		$5f^{n-1}6s^26p^66d^27s^2$	
An	Configuration	Term	Term	ΔE^a	Term	ΔE^b	Term^c	ΔE^d
Ac	$5f^06d^17s^2$	$^2D_{3/2}$			$^2D_{3/2}$	0.00		
Th	$5f^06d^27s^2$	$^3F_2$			$^3H_4$	1.39	$^3F_2$	0.00
Pa	$5f^26d^17s^2$	$^4K_{11/2}$			$^4K_{11/2}$	0.00	$^4H_{7/2}$	-0.25
U	$5f^36d^17s^2$	$^5L_6$			$^5L_6$	0.00	$^5L_6$	0.79
Np	$5f^46d^17s^2$	$^6L_{11/2}$			$^6L_{11/2}$	0.00	$^6M_{13/2}$	0.99
Pu	$5f^67s^2$	$^7F_0$	$^7F_0$	0.00	$^7K_4$	-0.87	$^7M_6$	0.57
Am	$5f^77s^2$	$^8S_{7/2}$	$^8S_{7/2}$	0.00	$^8H_{3/2}$	0.20		
Cm	$5f^76d^17s^2$	$^9D_2$	$^7F_6$	2.59	$^9D_2$	0.00		
Bk	$5f^97s^2$	$^6H_{15/2}$	$^6H_{15/2}$	0.00	$^8G_{13/2}$	-1.03		
Cf	$5f^{10}7s^2$	$^5I_8$	$^5I_8$	0.00	$(15/2,3/2)_8$	0.08		
Es	$5f^{11}7s^2$	$^4I_{15/2}$	$^4I_{15/2}$	0.00	$(8,3/2)_{15/2}$	0.38		
Fm	$5f^{12}7s^2$	$^3H_6$	$^3H_6$	0.00	$(15/2,3/2)_6$	0.54		
Md	$5f^{13}7s^2$	$^2F_{7/2}$	$^2F_{7/2}$	0.00	$(6,3/2)_{9/2}$	1.59		
No	$5f^{14}7s^2$	$^1S_0$	$^1S_0$	0.00	$(7/2,3/2)_2$	3.20		
Lr	$5f^{14}7s^27p^1$	$^2P_{1/2}$			$^2D_{3/2}$	0.66		

$^a \Delta E = E(5f^{n+1}6s^26p^67s^2) - E(\text{ground state})$.
$^b \Delta E = E(5f^n6s^26p^66d^17s^2) - E(\text{ground state})$.
^{c}Due to the large number of possible configurations the evaluation of some terms was not possible.
$^d \Delta E = E(5f^{n-1}6s^26p^66d^27s^2) - E(\text{ground state})$.

Table C.5: Experimentally observed ground states [71] and AE DHF energy differences (in eV) between these ground states and the lowest valence substate corresponding to pentavalent 5f-in-core LPPs ($5f^{n-2}$, n=2–6 for Pa–Am) and the hexavalent LPP for uranium ($5f^{n-3}$, n=3), respectively [30].

	Ground State		$5f^{n-2}6s^26p^66d^37s^2$		$5f^{n-3}6s^26p^66d^47s^2$	
An	Configuration	Term	Term^a	ΔE^b	Term	ΔE^c
Pa	$5f^26d^17s^2$	$^4K_{11/2}$	$^4F_{3/2}$	2.24		
U	$5f^36d^17s^2$	$^5L_6$	$^5I_4$	4.93	$^5D_0$	11.50
Np	$5f^46d^17s^2$	$^6L_{11/2}$				
Pu	$5f^67s^2$	$^7F_0$				
Am	$5f^77s^2$	$^8S_{7/2}$				

^{a}Due to the large number of possible configurations the evaluation of some terms was not possible.
$^b \Delta E = E(5f^{n-2}6s^26p^66d^37s^2) - E(\text{ground state})$.
$^c \Delta E = E(5f^{n-3}6s^26p^66d^47s^2) - E(\text{ground state})$.

Table C.6: Experimentally observed ground states [71] and experimental energy differences (in eV) between these ground states and the lowest valence substate corresponding to di- ($5f^{n+1}$, n=5–13 for Pu–No) [29], tri- ($5f^n$, n=0–14 for Ac–Lr) [54], and tetravalent ($5f^{n-1}$, n=1–9 for Th–Cf) [29] 5f-in-core LPPs, respectively. For the $6d^17s^2$ and $6d^27s^2$ valence sub-configurations of Fm–Lr and Am–Cf no experimental data are available, respectively.

An	Ground State Configuration	Term	$5f^{n+1}6s^26p^67s^2$ Term	ΔE^a	$5f^n6s^26p^66d^17s^2$ Term	ΔE^b	$5f^{n-1}6s^26p^66d^27s^2$ Term	ΔE^c
Ac	$5f^06d^17s^2$	$^2D_{3/2}$			$^2D_{3/2}$	0.00		
Th	$5f^06d^27s^2$	$^3F_2$			$^3H_4$	0.97	$^3F_2$	0.00
Pa	$5f^26d^17s^2$	$^4K_{11/2}$			$^4K_{11/2}$	0.00	$^4H_{7/2}$	0.25
U	$5f^36d^17s^2$	$^5L_6$			$^5L_6$	0.00	$^5L_6$	1.43
Np	$5f^46d^17s^2$	$^6L_{11/2}$			$^6L_{11/2}$	0.00	$^6M_{13/2}$	2.49
Pu	$5f^67s^2$	$^7F_0$	$^7F_0$	0.00	$^7K_4$	0.78	$^7M_6$	4.47
Am	$5f^77s^2$	$^8S_{7/2}$	$^8S_{7/2}$	0.00	$^8H_{3/2}$	1.32		
Cm	$5f^76d^17s^2$	$^9D_2$	$^7F_6$	0.15	$^9D_2$	0.00		
Bk	$5f^97s^2$	$^6H_{15/2}$	$^6H_{15/2}$	0.00	$^8G_{13/2}$	1.13		
Cf	$5f^{10}7s^2$	$^5I_8$	$^5I_8$	0.00	$(15/2,3/2)_8$	2.10		
Es	$5f^{11}7s^2$	$^4I_{15/2}$	$^4I_{15/2}$	0.00	$^6I_{17/2}$	2.40		
Fm	$5f^{12}7s^2$	$^3H_6$	$^3H_6$	0.00				
Md	$5f^{13}7s^2$	$^2F_{7/2}$	$^2F_{7/2}$	0.00				
No	$5f^{14}7s^2$	$^1S_0$	$^1S_0$	0.00				
Lr	$5f^{14}7s^27p^1$	$^2P_{1/2}$						

$^a \Delta E = E(5f^{n+1}6s^26p^67s^2) - E(\text{ground state})$.
$^b \Delta E = E(5f^n6s^26p^66d^17s^2) - E(\text{ground state})$.
$^c \Delta E = E(5f^{n-1}6s^26p^66d^27s^2) - E(\text{ground state})$.

Table C.7: An–F bond energies (in eV) $E_{bond} = [E(An) + 3 \times E(F) - E(AnF_3)]/3$ for AnF$_3$ (An=Ac–Lr) from LPP HF and SPP state-averaged MCSCF calculations [19, 20]. Additionally, the m.a.e. (in eV) as well as the m.r.e. with respect to the SPP data are given.

An	LPP	SPP
Ac	5.209	5.187
Th	5.081[a]	5.490
Pa	4.996[a]	5.252
U	4.939	5.052
Np	4.897	4.946
Pu	4.864	4.856
Am	4.840	4.772
Cm	4.823	4.718
Bk	4.803	4.720
Cf	4.794	4.795
Es	4.778	4.799
Fm	4.763	4.790
Md	4.757	4.819
No	4.746	4.804
Lr	4.734	4.768
m.a.e. (m.r.e.)	0.050 (1.0%)	

[a]For ThF$_3$ and PaF$_3$ the LPP 5f occupations are by up to 0.57 electrons larger than the SPP 5f occupations, because for these actinides the trivalent oxidation state is not preferred (Th) or even not stable (Pa) in aqueous solution [1] (cf. Sect. 3.1.5.6). Thus, for ThF$_3$ and PaF$_3$ the assumption of a near-integral 5f occupation is too crude [20], and the m.a.e. as well as m.r.e. were calculated neglecting the results for these systems.

Table C.8: Bond lengths R_e (in Å), angles $\angle$ (in deg), and total energies E (in a.u.) for the complexes [UO$_2$L$_2$] (L=sha, bha) from SPP DFT/B3LYP gas phase calculations. Both possible structures are given, i.e. the nitrogen atoms of the ligands located on the same (denoted as C_1) or on opposite sides (denoted as C_i).

	C_1-[UO$_2$sha$_2$]	C_i-[UO$_2$sha$_2$]	C_1-[UO$_2$bha$_2$]	C_i-[UO$_2$bha$_2$]
R_e(U–O$_{ax}$)	1.783	1.782	1.782	1.781
	1.783	1.782	1.782	1.781
R_e(U–O$_{Carb.}$)[a]	2.426	2.417	2.435	2.423
	2.427	2.416	2.438	2.423
R_e(U–ON)	2.313	2.322	2.315	2.324
	2.313	2.323	2.315	2.324
$\angle$O$_{ax}$–U–O$_{ax}$	175.5	180.0	175.4	180.0
$\angle$O$_{Carb.}$–U–ON[a]	66.6	66.7	66.9	66.8
	66.6	66.7	66.7	66.8
E	-1728.89381645	-1728.89378590	-1578.44002333	-1578.43980258

[a]O$_{Carb.}$ is the oxygen atom of the carbonyl group.

Table C.9: Experimentally observed ground states [71] and AE WB energy differences (in eV) between these ground states and the lowest states corresponding to tri- ($4f^n$, n=0–14 for La–Lu) [54] and tetravalent ($4f^{n-1}$, n=1–3, 8, 9 for Ce–Nd, Tb, Dy) [28] 4f-in-core LPPs, respectively.

Ln	Ground State Configuration	Term	$4f^n 5s^2 5p^6 5d^1 6s^2$ Term^a	ΔE^b	$4f^{n-1} 5s^2 5p^6 5d^2 6s^2$ Term	ΔE^c
La	$4f^0 5d^1 6s^2$	^2D	^2D	0.00		
Ce	$4f^1 5d^1 6s^2$	^1G	^1G	0.00	^3F	3.39
Pr	$4f^3 6s^2$	^4I	^4I		^4I	4.66
Nd	$4f^4 6s^2$	^5I	^5L	-0.21	^5L	5.59
Pm	$4f^5 6s^2$	^6H	^6L	-0.48		
Sm	$4f^6 6s^2$	^7F	^7F			
Eu	$4f^7 6s^2$	^8S	^8D			
Gd	$4f^7 5d^1 6s^2$	^9D	^9D	0.00		
Tb	$4f^9 6s^2$	^6H	^8G		^{10}F	-2.17
Dy	$4f^{10} 6s^2$	^5I	^7H		^9I	1.27
Ho	$4f^{11} 6s^2$	^4I	^6I			
Er	$4f^{12} 6s^2$	^3H	^5G			
Tm	$4f^{13} 6s^2$	^2F	^4K	-1.29		
Yb	$4f^{14} 6s^2$	^1S	^3H	0.44		
Lu	$4f^{14} 5d^1 6s^2$	^2D	^2D	0.00		

^{a}For Pr, Sm, Eu, Tb, Dy, Ho, and Er no AE WB energy corrections could be calculated, because there is more than one possibility to couple $4f^n$ and $5d^1$ to obtain the desired LS-states ^4I, ^7F, ^8D, ^8G, ^7H, ^6I, and ^5G, respectively.
$^b \Delta E = E(4f^n 5s^2 5p^6 5d^1 6s^2) - E$(ground state).
$^c \Delta E = E(4f^{n-1} 5s^2 5p^6 5d^2 6s^2) - E$(ground state).

Table C.10: Experimentally observed ground states [71] and AE DHF energy differences (in eV) between these ground states and the lowest states corresponding to tri- ($4f^n$, n=0–14 for La–Lu) [54] and tetravalent ($4f^{n-1}$, n=1–3, 8, 9 for Ce–Nd, Tb, Dy) [28] 4f-in-core LPPs, respectively.

Ln	Ground State Configuration	Ground State Term	$4f^n5s^25p^65d^16s^2$ Term	$4f^n5s^25p^65d^16s^2$ ΔE^a	$4f^{n-1}5s^25p^65d^26s^2$ Term	$4f^{n-1}5s^25p^65d^26s^2$ ΔE^b
La	$4f^05d^16s^2$	$^2D_{3/2}$	$^2D_{3/2}$	0.00		
Ce	$4f^15d^16s^2$	$^1G_4$	$^1G_4$	0.00	$^3F_2$	3.16
Pr	$4f^36s^2$	$^4I_{9/2}$	$^4I_{9/2}$	-0.27	$^4I_{7/2}$	4.21
Nd	$4f^46s^2$	$^5I_4$	$^5L_6$	-0.46	$^5L_6$	5.18
Pm	$4f^56s^2$	$^6H_{5/2}$	$^6L_{11/2}$	-0.76		
Sm	$4f^66s^2$	$^7F_0$	$^7F_4$	0.16		
Eu	$4f^76s^2$	$^8S_{7/2}$	$^8D_{3/2}$	1.49		
Gd	$4f^75d^16s^2$	$^9D_2$	$^9D_2$	0.00		
Tb	$4f^96s^2$	$^6H_{15/2}$	$^8G_{13/2}$	-2.96	$^{10}F_{3/2}$	-2.04
Dy	$4f^{10}6s^2$	$^5I_8$	$^7H_8$	-2.01	$^9I_7$	0.75
Ho	$4f^{11}6s^2$	$^4I_{15/2}$	$^6I_{17/2}$	-2.28		
Er	$4f^{12}6s^2$	$^3H_6$	$^5G_6$	-2.81		
Tm	$4f^{13}6s^2$	$^2F_{7/2}$	$(6,3/2)_{9/2}$	-2.07		
Yb	$4f^{14}6s^2$	$^1S_0$	$(7/2,3/2)_2$	-0.67		
Lu	$4f^{14}5d^16s^2$	$^2D_{3/2}$	$^2D_{3/2}$	0.00		

$^a\Delta E = E(4f^n5s^25p^65d^16s^2) - E(\text{ground state})$.
$^b\Delta E = E(4f^{n-1}5s^25p^65d^26s^2) - E(\text{ground state})$.

Table C.11: Experimentally observed ground states [71] and experimental energy differences (in eV) between these ground states and the lowest states corresponding to trivalent 4f-in-core LPPs ($4f^n$, n=0–14 for La–Lu) [54].

Ln	Ground State Configuration	Ground State Term	$4f^n5s^25p^65d^16s^2$ Term	$4f^n5s^25p^65d^16s^2$ $\Delta E^{a,b}$
La	$4f^05d^16s^2$	$^2D_{3/2}$	$^2D_{3/2}$	0.00
Ce	$4f^15d^16s^2$	$^1G_4$	$^1G_4$	0.00
Pr	$4f^36s^2$	$^4I_{9/2}$	$^4I_{9/2}$	0.55
Nd	$4f^46s^2$	$^5I_4$	$^5L_6$	0.84
Pm	$4f^56s^2$	$^6H_{5/2}$		
Sm	$4f^66s^2$	$^7F_0$	$^7F_0$	2.27
Eu	$4f^76s^2$	$^8S_{7/2}$	$^8D_{5/2}$	3.45
Gd	$4f^75d^16s^2$	$^9D_2$	$^9D_2$	0.00
Tb	$4f^96s^2$	$^6H_{15/2}$	$^8G_{13/2}$	0.04
Dy	$4f^{10}6s^2$	$^5I_8$	$^7H_8$	0.94
Ho	$4f^{11}6s^2$	$^4I_{15/2}$	$^6I_{17/2}$	1.04
Er	$4f^{12}6s^2$	$^3H_6$	$^5G_6$	0.89
Tm	$4f^{13}6s^2$	$^2F_{7/2}$	$(6,3/2)_{9/2}$	1.63
Yb	$4f^{14}6s^2$	$^1S_0$	$(7/2,3/2)_2$	2.88
Lu	$4f^{14}5d^16s^2$	$^2D_{3/2}$	$^2D_{3/2}$	0.00

$^a\Delta E = E(4f^n5s^25p^65d^16s^2) - E(\text{ground state})$.
^{b}Energy differences from [43].

Table C.12: Atomization energies ΔE_{at} (in eV) for LnF$_3$ (Ln=La–Lu) with respect to the valence substates 4f^{n}5d^{1}6s^2 from LPP HF and SPP state-averaged MCSCF calculations. SPP values are given for the lowest LS-states according to Hund's rule, while LPP values are given for ^2D states, since here the 4f shell is treated in an averaged manner. Additionally, LPP atomization energies corrected to account for the coupling between 4f^n and 5d^1 $\Delta E_{coupl.}$ (in eV), and LPP as well as SPP values corrected to account for the proper description of triply-charged ions Ln^{3+} $\Delta E_{coupl.+Ln^{3+}}$ and $\Delta E_{at+Ln^{3+}}$ (in eV) are given, respectively [54].

Ln	ΔE_{at}			$\Delta E_{coupl.}$		$\Delta E_{coupl.+Ln^{3+}}$	$\Delta E_{at+Ln^{3+}}$
	LPP	SPP[a]		LPP		LPP	SPP
La	15.83	^2D	15.95	^2D	15.83	15.78	15.95
Ce	15.80	^1G	15.18	^1G	15.13	15.02	15.34
Pr	15.76	^4K	15.32[b]	^4K	15.30	15.18	15.55
Nd	15.73	^5L	15.08[b]	^5L	15.12	15.00	15.31
Pm	15.73	^6L	14.88	^6L	14.94	14.80	15.06
Sm	15.72	^7K	14.90	^7K	14.81	14.67	15.01
Eu	15.73	^8H	15.00	^8H	14.86	14.73	15.07
Gd	15.69	^9D	14.85	^9D	15.12	15.02	15.22
Tb	15.73	^8H	15.74	^8H	15.51	15.41	15.68
Dy	15.75	^7K	15.78	^7K	15.60	15.52	15.77
Ho	15.78	^6L	15.69	^6L	15.57	15.50	15.63
Er	15.82	^5L	15.67	^5L	15.50	15.44	15.55
Tm	15.84	^4K	15.80	^4K	15.46	15.41	15.59
Yb	15.89	^3H	16.16	^3H	15.58	15.54	15.72
Lu	15.81	^2D	15.87	^2D	15.81	15.78	15.99

[a]In some cases the calculated LS-state does not correspond to the lowest LS-state at the HF level, i.e. $^8G/^8F/^8D$, $^7H/^7I/^7G/^7F$, $^6I/^6K/^6H$, $^5G/^5H$, ^4F, and ^3P are lower for Tb, Dy, Ho, Er, Tm, and Yb, respectively.

[b]Since for Pr and Nd $^4K/^4I$ and $^5L/^5K$ states are nearly degenerate, respectively, the energies corresponding to ^4K and ^5L were taken from SPP state-averaged MCSCF calculations, where these degenerate states were calculated simultaneously.

Table C.13: Ionic binding energies ΔE_{ion} (in eV) for LnF$_3$ (Ln=La–Lu) from LPP+CPP CCSD(T) calculations [54].

Ln	ΔE_{ion}
La	45.41
Ce	45.77
Pr	46.12
Nd	46.46
Pm	46.78
Sm	47.09
Eu	47.41
Gd	47.71
Tb	48.03
Dy	48.35
Ho	48.67
Er	49.00
Tm	49.30
Yb	49.63
Lu	49.80

Bibliography

[1] N. Kaltsoyannis, P. Scott, *The f Elements*, 1. Ed., Oxford University Press, Oxford (1999).

[2] A. F. Holleman, E. Wiberg, N. Wiberg, *Lehrbuch der Anorganischen Chemie*, 101. Ed., Walter de Gruyter, Berlin (1995).

[3] J. J. Katz, G. T. Seaborg, L. R. Morss, *The Chemistry of the Actinide Elements*, Vol. 2, 2. Ed., Chapman and Hall, London (1986).

[4] http://www.fzd.de/pls/rois/Cms?pOid=14476.

[5] M. Pepper, B. E. Bursten, Chem. Rev. **91**, 719 (1991).

[6] G. Schreckenbach, P. J. Hay, R. L. Martin, J. Comp. Chem. **20**, 70 (1999).

[7] M. Dolg, X. Cao, in: *Recent Advances in Relativistic Molecular Theory* (Eds. K. Hirao, Y. Ishikawa), Vol. 5, Chap. 1, World Scientific, New Jersey (2003).

[8] X. Cao, M. Dolg, Coord. Chem. Rev. **250**, 900 (2006).

[9] M. Dolg, H. Stoll, in: *Handbook on the Physics and Chemistry of Rare Earths* (Eds. K. A. Gschneidner Jr., L. Eyring), Vol. 22, Chap. 152, Elsevier Science B.V., Amsterdam (1996).

[10] X. Cao, M. Dolg, in: *Relativistic Methods for Chemists* (Eds. M. Barysz, Y. Ishikawa), Chap. 1, Springer (2009), submitted.

[11] S. Huzinaga, L. Seijo, Z. Barandiarán, M. Klobukowski, J. Chem. Phys. **86**, 2132 (1987).

[12] P. J. Hay, P. J. Wadt, J. Chem. Phys. **82**, 270 (1985).

[13] P. J. Hay, R. L. Martin, J. Chem. Phys. **109**, 3875 (1998).

[14] M. Dolg, H. Stoll, H. Preuss, J. Chem. Phys. **90**, 1730 (1989).

[15] W. Küchle, M. Dolg, H. Stoll, H. Preuss, J. Chem. Phys. **100**, 7535 (1994).

[16] M. Dolg, X. Cao, J. Phys. Chem. A (2009), accepted.

[17] M. Dolg, H. Stoll, A. Savin, H. Preuss, Theor. Chim. Acta **75**, 173 (1989).

[18] M. Dolg, H. Stoll, H. Preuss, Theor. Chim. Acta **85**, 441 (1993).

[19] A. Moritz, *Relativistische Energie-konsistente Pseudopotentiale für Actinoiden mit Fester 5f-Besetzung*, diploma thesis, Universität zu Köln (2005).

[20] A. Moritz, X. Cao, M. Dolg, Theor. Chem. Acc. **117**, 473 (2007).

[21] X. Cao, M. Dolg, H. Stoll, J. Chem. Phys. **118**, 487 (2003).

[22] X. Cao, M. Dolg, Mol. Phys. **101**, 2427 (2003).

[23] X. Cao, Q. Li, A. Moritz, Z. Xie, M. Dolg, X. Chen, W. Fang, Inorg. Chem. **45**, 3444 (2006).

[24] J. Wiebke, A. Moritz, X. Cao, M. Dolg, Phys. Chem. Chem. Phys. **9**, 459 (2007).

[25] R. A. Evarestov, M. V. Losev, A. I. Panin, N. S. Mosyagin, A. V. Titov, Phys. Stat. Sol. (b) **245**, 114 (2008).

[26] M. Dolg, in: *Relativistic Electronic Structure Theory, Part 1: Fundamentals* (Ed. P. Schwerdtfeger), Vol. 11, Chap. 14, Elsevier, Amsterdam (2002).

[27] Y.-K. Han, K. Hirao, Chem. Phys. Lett. **324**, 453 (2000).

[28] M. Hülsen, A. Weigand, M. Dolg, Theor. Chem. Acc. **122**, 23 (2009).

[29] A. Moritz, X. Cao, M. Dolg, Theor. Chem. Acc. **118**, 845 (2007).

[30] A. Moritz, M. Dolg, Theor. Chem. Acc. **121**, 297 (2008).

[31] H. Partridge, in: *Encyclopedia of Computational Chemistry* (Eds. P. von Ragué Schleyer *et al.*), Vol. 1, 581, Wiley & Sons, New York (1998).

[32] W. Müller, J. Flesch, W. Meyer, J. Chem. Phys. **80**, 3297 (1984).

[33] W. Müller, W. Meyer, J. Chem. Phys. **80**, 3311 (1984).

[34] P. Fuentealba, H. Preuss, H. Stoll, L. von Szentpály, Chem. Phys. Lett. **89**, 418 (1982).

[35] P. Fuentealba, J. Phys. B: At. Mol. Opt. Phys. **15**, L555 (1982).

[36] L. von Szentpály, P. Fuentealba, H. Preuss, H. Stoll, Chem. Phys. Lett. **93**, 555 (1982).

[37] J. Flad, G. Igel, M. Dolg, H. Stoll, H. Preuss, Chem. Phys. **75**, 331 (1983).

[38] A. Savin, M. Dolg, H. Stoll, H. Preuss, J. Flesch, Chem. Phys. Lett. **100**, 455 (1983).

[39] H. Stoll, P. Fuentealba, M. Dolg, J. Flad, L. von Szentpály, H. Preuss, J. Chem. Phys. **79**, 5532 (1983).

[40] G. Igel, U. Wedig, M. Dolg, P. Fuentealba, H. Preuss, H. Stoll, R. Frey, J. Chem. Phys. **81**, 2737 (1984).

[41] Y. Wang, M. Dolg, Theor. Chem. Acc. **100**, 124 (1998).

[42] I. N. Levine, *Quantum Chemistry*, 5. Ed., Prentice-Hall, New Jersey (2000).

[43] M. Dolg, *Energiejustierte Quasirelativistische Pseudopotentiale für die 4f-Elemente*, PhD thesis, Universität Stuttgart (1989).

[44] K. G. Dyall, I. P. Grant, C. T. Johnson, F. A. Parpia, E. P. Plummer, Comput. Phys. Commun. **55**, 425 (1989), atomic structure code GRASP; extension for pseudopotentials by M. Dolg and B. Metz.

[45] D. Figgen, *Relativistische Pseudopotentiale: Multikonfigurations-Dirac-Hartree-Fock-Justierung für 4d- und 5d-Elemente und Anwendung in Molekülrechnungen mit Spin-Bahn-Kopplung*, PhD thesis, Universität Stuttgart (2007).

[46] J. H. Wood, A. M. Boring, Phys. Rev. B **18**, 2701 (1978).

[47] M. Dolg, Theor. Chem. Acc. **114**, 297 (2005).

[48] K. A. Peterson, D. Figgen, M. Dolg, H. Stoll, J. Chem. Phys. **126**, 124101 (2007).

[49] P. Hohenberg, W. Kohn, Phys. Rev. B **136**, 864 (1964).

[50] P. Atkins, R. Friedman, *Molecular Quantum Mechanics*, 4. Ed., Oxford University Press (2005).

[51] J. Reinhold, *Quantentheorie der Moleküle*, Teubner, Stuttgart (1994).

[52] H.-J. Werner, P.J. Knowles, R. Lindh, F.R. Manby, M. Schütz, *et al.*, *Molpro, version 2006.1, a package of ab initio programs*, tech. rep., University of Birmingham (2006).

[53] W. Kohn, L. J. Sham, Phys. Rev. A **140**, 1133 (1965).

[54] A. Weigand, X. Cao, J. Yang, M. Dolg, Theor. Chem. Acc. **126**, 117 (2010).

[55] C. Froese-Fischer, *The Hartree–Fock Method for Atoms.*, Wiley, New York (1977); program MCHF77, modified for pseudopotentials and quasirelativistic calculations by M. Dolg (1995).

[56] D. Kolb, W. R. Johnson, P. Shorer, Phys. Rev. A **26**, 19 (1982).

[57] R. M. Pitzer, *Atomic electronic structure code ATMSCF.*, tech. rep., University of Ohio State (1979).

[58] F. Weigend, R. Ahlrichs, Phys. Chem. Chem. Phys. **7**, 3297 (2005).

[59] J. Blaise, J. Wyart, in: *International Tables of Selected Constants*, Vol. 20, CNRS, Paris (1992).

[60] S. Köhler, R. Deissenberger, K. Eberhardt, *et al.*, Spectrochim. Acta, Part B **52**, 717 (1997).

[61] J. R. Peterson, N. Erdmann, M. Nunnemann, *et al.*, J. Alloys Compd. **271**, 876 (1998).

[62] T. H. Dunning Jr., J. Chem. Phys. **90**, 1007 (1989).

[63] R. A. Kendall, T. H. Dunning Jr., R. J. Harrison, J. Chem. Phys. **96**, 6769 (1992).

[64] G. V. Girichev, K. V. Krasnov, N. I. Giricheva, O. G. Krasnova, J. Struct. Chem. **40**, 207 (1999).

[65] R. J. M. Konings, D. L. Hildenbrand, J. Alloys Compd. **271**, 583 (1998).

[66] L. H. Jones, S. Ekberg, J. Chem. Phys. **67**, 2591 (1977).

[67] M. Kimura, V. Schomaker, D. W. Smith, J. Chem. Phys. **48**, 4001 (1968).

[68] H. M. Seip, Acta Chem. Scand. **19**, 1955 (1965).

[69] A. Kovacs, R. J. M. Konings, D. S. Nemcsok, J. Alloys Compd. **353**, 128 (2003).

[70] M. Dolg, H. Stoll, Theor. Chim. Acta **75**, 369 (1989).

[71] http://www.physics.nist.gov/PhysRefData/Handbook/element_name.htm.

[72] S. G. Wang, W. H. E. Schwarz, J. Phys. Chem. **99**, 11687 (1995).

[73] Y. Mochizuki, H. Tatewaki, J. Chem. Phys. **118**, 9201 (2003).

[74] Z. Akdeniz, A. Karaman, M. P. Tosi, Z. Naturforsch. **56a**, 376 (2001).

[75] W. R. Wadt, P. J. Hay, J. Am. Chem. Soc. **101**, 5198 (1979).

[76] E. R. Batista, R. L. Martin, P. J. Hay, J. Chem. Phys. **121**, 11104 (2004).

[77] Y. K. Han, J. Comput. Chem. **22**, 2010 (2001).

[78] E. R. Batista, R. L. Martin, P. J. Hay, J. E. Peralta, G. E. Scuseria, J. Chem. Phys. **121**, 2144 (2004).

[79] L. Joubert, P. Maldivi, J. Phys. Chem. A **105**, 9068 (2001).

[80] V. Vetere, B. O. Roos, P. Maldivi, C. Adamo, Chem. Phys. Lett. **396**, 452 (2004).

[81] R. W. Field, Ber. Bunsenges. Phys. Chem. **86**, 771 (1982).

[82] Y. Wang, F. Schautz, H. J. Flad, M. Dolg, J. Phys. Chem. A **103**, 5091 (1999).

[83] A. Moritz, M. Dolg, Chem. Phys. **337**, 48 (2007).

[84] J. Wiebke, A. Weigand, D. Weißmann, M. Glorius, H. Moll, G. Bernhard, M. Dolg, Inorg. Chem. (2010), accepted.

[85] C. Park, J. Almlöf, J. Chem. Phys. **95**, 1829 (1991).

[86] A. Streitwieser Jr., U. Müller-Westerhoff, J. Am. Chem. Soc. **90**, 7364 (1968).

[87] R. D. Fischer, Theor. Chim. Acta **1**, 418 (1963).

[88] A. Streitwieser Jr., N. Yoshida, J. Am. Chem. Soc. **91**, 7528 (1969).

[89] J. Goffart, J. Fuger, D. Brown, G. Duyckaerts, Inorg. Nucl. Chem. Lett. **10**, 413 (1974).

[90] D. G. Karraker, J. A. Stone, E. R. Jones Jr., N. Edelstein, J. Am. Chem. Soc. **92**, 4841 (1970).

[91] A. Avdeef, K. N. Raymond, K. O. Hodgson, A. Zalkin, Inorg. Chem. **11**, 1083 (1972).

[92] N. Rösch, A. Streitwieser Jr., J. Organomet. Chem. **145**, 195 (1978).

[93] N. Rösch, A. Streitwieser Jr., J. Am. Chem. Soc. **105**, 7237 (1983).

[94] J. G. Brennan, J. C. Green, C. M. Redfern, J. Am. Chem. Soc. **111**, 2373 (1989).

[95] A. H. H. Chang, R. M. Pitzer, J. Am. Chem. Soc. **111**, 2500 (1989).

[96] M. Dolg, P. Fulde, H. Stoll, H. Preuss, A. Chang, R. M. Pitzer, Chem. Phys. **195**, 71 (1995).

[97] P. M. Boerrigter, E. J. Baerends, J. G. Snijders, Chem. Phys. **122**, 357 (1988).

[98] W. Liu, M. Dolg, P. Fulde, J. Chem. Phys. **107**, 3584 (1997).

[99] R. Ahlrichs, M. Bär, H. P. Baron, R. Bauernschmitt, S. Böcker, M. Ehrig, K. Eichkorn, S. Elliott, F. Furche, F. Haase, M. Häser, H. Horn, C. Huber, U. Huniar, C. Kölmel, M. Kollwitz, C. Ochsenfeld, H. Öhm, A. Schäfer, U. Schneider, O. Treutler, M. Arnim, F. Weigend, P. Weis, H. Weiss, *Turbomole 5.7*, tech. rep., Universität Karlsruhe (2004).

[100] A. D. Becke, J. Chem. Phys. **98**, 5648 (1993).

[101] C. T. Lee, W. T. Yang, R. G. Parr, Phys. Rev. B: Condens. Matter Mater. Phys. **37**, 785 (1988).

[102] A. D. Becke, Phys. Rev. A: At., Mol., Opt. Phys. **38**, 3098 (1988).

[103] S. H. Vosko, L. Wilk, M. Nusair, Can. J. Phys. **58**, 1200 (1980).

[104] J. C. Slater, Phys. Rev. **81**, 385 (1951).

[105] P. A. M. Dirac, Proc. Cambr. Phil. Soc. **26**, 376 (1930).

[106] J. P. Perdew, Y. Wang, Phys. Rev. B: Condens. Matter Mater. Phys. **45**, 13244 (1992).

[107] J. Li, B. E. Bursten, J. Am. Chem. Soc. **120**, 11456 (1998).

[108] N. Kaltsoyannis, Inorg. Chem. **39**, 6009 (2000).

[109] R. G. Denning, J. Phys. Chem. A **111**, 4125 (2007).

[110] X. Cao, M. Dolg, J. Mol. Struct. **673**, 203 (2004).

[111] K. G. Dyall, Mol. Phys. **96**, 511 (1999).

[112] H. H. Cornehl, C. Heinemann, J. Marcalo, A. P. de Matos, H. Schwarz, Angew. Chem., Int. Ed. **35**, 891 (1996).

[113] F. Real, V. Vallet, C. Marian, U. Wahlgren, J. Chem. Phys. **127**, 214302 (2007).

[114] W. A. de Jong, L. Visscher, W. C. Nieuwpoort, J. Mol. Struct. **458**, 41 (1999).

[115] G. A. Shamov, G. Schreckenbach, R. L. Martin, P. J. Hay, Inorg. Chem. **47**, 1465 (2008).

[116] V. Vallet, L. Maron, B. Schimmelpfennig, T. Leininger, C. Teichteil, O. Gropen, I. Grenthe, U. Wahlgren, J. Phys. Chem. A **103**, 9285 (1999).

[117] N. V. Jarvis, R. D. Hancock, Inorg. Chim. Acta **182**, 229 (1991).

[118] J. R. Brainard, B. A. Strietelmeier, P. H. Smith, P. J. Langstonunkefer, M. E. Barr, R. R. Ryan, Radiochim. Acta **58-9**, 357 (1992).

[119] A.-M. Albrecht-Gary, S. Blanc, N. Rochel, A. Z. Ocaktan, M. A. Abdallah, Inorg. Chem. **33**, 6391 (1994).

[120] M. Bouby, I. Billard, J. MacCordick, J. Alloys Compd. **271**, 206 (1998).

[121] M. Bouby, I. Billard, J. MacCordick, Czech. J. Phys. **49**, 769 (1999).

[122] M. P. Neu, J. H. Matonic, C. E. Ruggiero, B. L. Scott, Angew. Chem. Int. Ed. **39**, 1442 (2000).

[123] H. Moll, M. Glorius, G. Bernhard, A. Johnsson, K. Pedersen, M. Schäfer, H. Budzikiewicz, Geomicrobio. J. **25**, 157 (2008).

[124] H. Moll, A. Johnsson, M. Schäfer, K. Pedersen, H. Budzikiewicz, G. Bernhard, BioMetals **21**, 219 (2008).

[125] H. Budzikiewicz, Fortschr. Chem. Org. Naturst. **87**, 83 (2004).

[126] M. Glorius, H. Moll, G. Bernhard, Radiochim. Acta **95**, 151 (2007).

[127] M. Glorius, H. Moll, G. Geipel, G. Bernhard, J. Radioanal. Nucl. Chem. **277**, 371 (2008).

[128] S. Gez, R. Luxenhofer, A. Levina, R. Codd, P. A. Lay, Inorg. Chem. **44**, 2934 (2005).

[129] M. D. Hall, T. W. Failes, D. E. Hibbs, T. W. Hambley, Inorg. Chem. **41**, 1223 (2002).

[130] D. Vulpius, in: *Zur Komplexbildung von Actiniden (U, Np) mit Hydroxybenzoesäuren*, Dresden University of Technology (2005).

[131] A. W. H. Lam, W. T. Wong, S. Gao, G. H. Wen, X. X. Zhang, Eur. J. Inorg. Chem. (1), 149 (2003).

[132] B. Kurzak, H. Kozlowski, E. Farkas, Coord. Chem. Rev. **114**, 169 (1992).

[133] K. E. Gutowski, V. A. Cocalia, S. T. Griffin, N. J. Bridges, D. A. Dixon, R. D. Rogers, J. Am. Chem. Soc. **129**, 526 (2007).

[134] A. Ikeda, C. Hennig, S. Tsushima, K. Takao, Y. Ikeda, A. C. Scheinost, G. Bernhard, Inorg. Chem. **46**, 4212 (2007).

[135] C. Gaillard, A. Chaumont, C. Billard, C. Hennig, A. Ouadi, G. Wipff, Inorg. Chem. **46**, 4815 (2007).

[136] G. A. Shamov, G. Schreckenbach, J. Phys. Chem. A **110**, 9486 (2006).

[137] Z. Szabo, T. Toraishi, V. Vallet, I. Grenthe, Coord. Chem. Rev. **250**, 784 (2006).

[138] V. Vallet, Z. Szabo, I. Grenthe, Dalton Trans. (22), 3799 (2004).

[139] V. Vallet, H. Moll, U. Wahlgren, Z. Szabo, I. Grenthe, Inorg. Chem. **42**, 1982 (2003).

[140] V. Vallet, U. Wahlgren, B. Schimmelpfennig, H. Moll, Z. Szabo, I. Grenthe, Inorg. Chem. **40**, 3516 (2001).

[141] U. Wahlgren, H. Moll, I. Grenthe, B. Schimmelpfennig, L. Maron, V. Vallet, O. Gropen, J. Phys. Chem. A **103**, 8257 (1999).

[142] R. Bauernschmitt, R. Ahlrichs, Chem. Phys. Lett. **256**, 454 (1996).

[143] R. Bauernschmitt, M. Häser, O. Treutler, R. Ahlrichs, Chem. Phys. Lett. **264**, 573 (1997).

[144] J. Wiebke, A. Moritz, M. Glorius, H. Moll, G. Bernhard, M. Dolg, Inorg. Chem. **47**, 3150 (2008).

[145] E. R. Batista, R. L. Martin, P. J. Hay, J. E. Peralta, G. E. Scuseria, J. Chem. Phys. **121**, 2144 (2004).

[146] J. Tomasi, B. Mennucci, R. Cammi, Chem. Rev. **105**, 2999 (2005).

[147] K. I. M. Ingram, J. L. Haller, N. Kaltsoyannis, Dalton Trans. **20**, 2403 (2006).

[148] K. E. Gutowski, D. A. Dixon, J. Phys. Chem. A **110**, 8840 (2006).

[149] G. A. Shamov, G. Schreckenbach, J. Phys. Chem. A **109**, 10961 (2005).

[150] L. V. Moskaleva, S. Krüger, A. Spörl, N. Rösch, Inorg. Chem. **43**, 4080 (2004).

[151] J. Haller, N. Kaltsoyannis, in: *Recent Advances in Actinide Science* (Eds. R. Alvarez, N. D. Bryan, I. May), RSC Publishing, Cambridge (2006).

[152] A. Klamt, G. Schüürmann, J. Chem. Soc. Perkin Trans. **2**, 799 (1993).

[153] A. Schäfer, A. Klamt, D. Sattel, J. C. W. Lohrenz, F. Eckert, Phys. Chem. Chem. Phys. **2**, 2187 (2000).

[154] A. Klamt, V. Jonas, T. Bürger, J. C. W. Lohrenz, J. Phys. Chem. A **102**, 5074 (1998).

[155] Y. Mochizuki, S. Tsushima, Chem. Phys. Lett. **372**, 114 (2003).

[156] T. Yang, B. E. Bursten, Inorg. Chem. **45**, 5291 (2006).

[157] Z. Cao, K. Balasubramanian, J. Chem. Phys. **123**, 114309 (2005).

[158] K. Balasubramanian, D. Chaudhuri, Chem. Phys. Lett. **450**, 196 (2008).

[159] J. Neugebauer, B. A. Hess, J. Chem. Phys. **118**, 7215 (2003).

[160] S. Matsika, Z. Zhang, S. R. Brozell, J. P. Blaudeau, Q. Wang, R. M. Pitzer, J. Phys. Chem. A **105**, 3825 (2001).

[161] Z. Zhang, R. M. Pitzer, J. Phys. Chem. A **103**, 6880 (1999).

[162] K. Pierloot, E. van Besien, J. Chem. Phys. **123**, 204309 (2005).

[163] F. Real, V. Vallet, C. Marian, U. Wahlgren, J. Chem. Phys. **127**, 214302 (2007).

[164] J. P. Perdew, Phys. Rev. B **33**, 8822 (1986).

[165] J. P. Perdew, K. B. M. Ernzerhof, Phys. Rev. Lett. **77**, 3865 (1996).

[166] J. P. Perdew, K. B. M. Ernzerhof, K. Burke, J. Chem. Phys. **105**, 9982 (1996).

[167] A. Dreuw, J. L. Weisman, M. Head-Gordon, J. Chem. Phys. **119**, 2943 (2003).

[168] J. Yang, M. Dolg, Theor. Chem. Acc. **113**, 212 (2005).

[169] N. I. Giricheva, O. G. Krasnova, G. V. Girichev, J. Struct. Chem. **39**, 192 (1998).

[170] G. Lanza, I. L. Fragala, J. Phys. Chem. A **102**, 7990 (1998).

[171] X. Cao, M. Dolg, J. Chem. Phys. **115**, 7348 (2001).

[172] C. Heinemann, H. H. Cornehl, D. Schröder, M. Dolg, H. Schwarz, Inorg. Chem. **35**, 2463 (1996).

[173] A. Kerridge, R. Coates, N. Kaltsoyannis, J. Phys. Chem. A **113**, 2896 (2009).

[174] S. D. Gabelnick, G. T. Reedy, M. G. Chasanov, J. Chem. Phys. **60**, 1167 (1974).

[175] M. Hargittai, Coord. Chem. Rev. **91**, 35 (1988).

[176] C. E. Myers, Inorg. Chem. **14**, 199 (1975).

[177] P. D. Kleinschmidt, K. H. Lau, D. L. Hildenbrand, J. Chem. Phys. **74**, 653 (1981).

[178] L. Joubert, G. Picard, J. J. Legendre, Inorg. Chem. **37**, 1984 (1998).

[179] G. Lanza, Z. Varga, M. Kolonits, M. Hargittai, J. Chem. Phys. **128**, 074301 (2008).

[180] T. R. Cundari, S. O. Sommerer, L. A. Strohecker, L. Tippett, J. Chem. Phys. **103**, 7058 (1995).

[181] C. Clavaguera, J.-P. Dognon, P. Pyykkö, Chem. Phys. Lett. **429**, 8 (2006).

[182] N. I. Popenko, E. Z. Zasorin, V. P. Spiridonov, A. A. Ivanov, Inorg. Chim. Acta **31**, L371 (1978).

[183] X. Cao, M. Dolg, J. Mol. Struct. **581**, 139 (2002).

[184] M. J. Frisch, G. W. Trucks, H. B. Schlegel, G. E. Scuseria, M. A. Robb, J. R. Cheeseman, J. A. M. Jr., T. Vreven, K. N. Kudin, J. C. Burant, J. M. Millam, S. S. Iyengar, J. Tomasi, V. Barone, B. Mennucci, M. Cossi, G. Scalmani, N. Rega, G. A. Petersson, H. Nakatsuji, M. Hada, M. Ehara, K. T. R. Fukuda, J. Hasegawa, M. Ishida, T. Nakajima, Y. Honda, O. Kitao, H. Nakai, M. Klene, X. Li, J. E. Knox, H. P. Hratchian, J. B. Cross, V. Bakken, C. Adamo, J. Jaramillo, R. Gomperts, R. E. Stratmann, O. Yazyev, A. J. Austin, R. Cammi, C. Pomelli, J. W. Ochterski, P. Y. Ayala, K. Morokuma, G. A. Voth, P. Salvador, J. J. Dannenberg, V. G. Zakrzewski, S. Dapprich, A. D. Daniels, M. C. Strain, O. Farkas, D. K. Malick, A. D. Rabuck, K. Raghavachari, J. B. Foresman, J. V. Ortiz, Q. Cui, A. G. Baboul, S. Clifford, J. Cioslowski, B. B. Stefanov, G. Liu, A. Liashenko, P. Piskorz, I. Komaromi, R. L. Martin, D. J. Fox, T. Keith, M. A. Al-Laham, C. Y. Peng, A. Nanayakkara, M. Challacombe, P. M. W. Gill, B. Johnson, W. Chen, M. W. Wong, C. Gonzalez, J. A. Pople, *Gaussian 03, Revision D.02*, tech. rep., Inc., Wallingford CT (2004).

[185] B. O. Roos, R. Lindh, P.-A. Malmqvist, V. Veryazov, P.-O. Widmark, A. C. Borin, J. Phys. Chem. A **112**, 11431 (2008).

[186] M. Dolg, H. Stoll, H. Preuss, J. Mol. Struct. **235**, 67 (1991).

[187] A. Lesar, G. Muri, M. Hodoscek, J. Chem. Phys. **102**, 1170 (1998).

[188] A. Weigand, X. Cao, V. Vallet, J.-P. Flament, M. Dolg, J. Phys. Chem. A **113**, 11509 (2009).

[189] C. Danilo, V. Vallet, J.-P. Flament, U. Wahlgren, J. Chem. Phys. **128**, 154310 (2008).

[190] V. Kaufman, L. F. Radziemski Jr., J. Opt. Soc. Am. **66**, 599 (1976).

[191] J. F. Wyart, V. Kaufman, J. Sugar, Phys. Scr. **22**, 389 (1980).

[192] C. H. H. van Deurzen, K. Rajnak, J. G. Conway, J. Opt. Soc. Am. B **1**, 45 (1984).

[193] I. Infante, E. Eliav, M. J. Vilkas, Y. Ishikawa, U. Kaldor, L. Visscher, J. Chem. Phys. **127**, 124308 (2007).

[194] Z. Barandiarán, L. Seijo, J. Chem. Phys. **118**, 7439 (2003).

[195] M. Seth, K. G. Dyall, R. Shepard, A. Wagner, J. Phys. B: At. Mol. Opt. Phys. **34**, 2383 (2001).

[196] E. Eliav, U. Kaldor, Y. Ishikawa, Phys. Rev. A **51**, 225 (1995).

[197] E. Eliav, U. Kaldor, Y. Ishikawa, Phys. Rev. A **49**, 1724 (1994).

[198] A. Landau, E. Eliav, Y. Ishikawa, U. Kaldor, J. Chem. Phys. **115**, 6862 (2001).

[199] G. Karlström, R. Lindh, P. A. Malmqvist, B. O. Roos, U. Ryde, V. Veryazov, P. O. Widmark, M. Cossi, B. Schimmelpfennig, P. Neogrady, L. Seijo, *MOLCAS, a program package for computational chemistry*, tech. rep. (2003).

[200] V. Vallet, L. Maron, C. Teichteil, J.-P. Flament, J. Chem. Phys. **113**, 1391 (2000).

[201] R. Llusar, M. Casarrubios, Z. Barandiarán, L. Seijo, J. Chem. Phys. **105**, 5321 (1996).

[202] E. Eliav, M. J. Vilkas, Y. Ishikawa, U. Kaldor, J. Chem. Phys. **122**, 224113 (2005).

[203] B. Schimmelpfennig, *An atomic mean-field integral program*, tech. rep. (1996).

[204] Z. B. Goldschmidt, Phys. Rev. A **27**, 740 (1983).

Danksagung

Prof. Dr. Michael Dolg danke ich sehr herzlich für die Möglichkeit, im Bereich der Pseudopotentiale zu promovieren, für die gute Betreuung und die stete Unterstützung bei fachlichen Problemen.

Prof. Dr. Ulrich Deiters danke ich für die Begutachtung meiner Dissertation und Prof. Dr. Uwe Ruschewitz für das Übernehmen des Prüfungsvorsitzes meiner Disputation.

Dr. Xiaoyan Cao danke ich sehr herzlich für die vielen Diskussionen, das Beantworten von all meinen Fragen, die gemeinsame Fahrt nach Lille zum Erlernen des Programms EPCISO...

Jonas Wiebke, Maja Glorius, Daniel Weißmann, Dr. Henry Moll, Dr. Valérie Vallet und Dr. Jean-Pierre Flament danke ich für die gute Zusammenarbeit auf dem Gebiet der Uranyl(VI)-Komplexe bzw. der U^{5+}/U^{4+}-Feinstrukturaufspaltung.

Michael Hülsen danke ich für die guten Ergebnisse während seines Praktikums, welche ich in meiner Doktorarbeit verwendet habe.

Allen Mitarbeitern und ehemaligen Mitarbeitern des Arbeitskreises Theoretische Chemie danke ich für die freundliche und kollegiale Atmosphäre, die schönen Weihnachts- und Promotionsfeiern, die Grillabende...

Der Deutschen Forschungsgemeinschaft danke ich für die finanzielle Unterstützung.

Meinen Studienkollegen, ohne die vor allem die Praktika sehr viel schwieriger zu ertragen gewesen wären, möchte ich recht herzlich für die schöne Zeit und die Freundschaft danken.

Meinen Eltern und meinen Schwestern danke ich für einfach alles. Ohne Eure Unterstüzung wäre mir dieses Studium nicht möglich gewesen.

Ganz besonders möchte ich meinem Mann Holger danken, der immer für mich da ist und mich so liebt, wie ich bin.

Die VDM Verlagsservicegesellschaft sucht für wissenschaftliche Verlage abgeschlossene und herausragende

Dissertationen, Habilitationen, Diplomarbeiten, Master Theses, Magisterarbeiten usw.

für die kostenlose Publikation als Fachbuch.

Sie verfügen über eine Arbeit, die hohen inhaltlichen und formalen Ansprüchen genügt, und haben Interesse an einer honorarvergüteten Publikation?

Dann senden Sie bitte erste Informationen über sich und Ihre Arbeit per Email an *info@vdm-vsg.de*.

Sie erhalten kurzfristig unser Feedback!

VDM Verlagsservicegesellschaft mbH
Dudweiler Landstr. 99   Telefon +49 681 3720 174
D - 66123 Saarbrücken   Fax   +49 681 3720 1749

www.vdm-vsg.de

Die VDM Verlagsservicegesellschaft mbH vertritt

MIX
Papier aus verantwortungsvollen Quellen
Paper from responsible sources
FSC® C105338

Printed by Books on Demand GmbH, Norderstedt / Germany